Remains of ancient raised fields, possibly 2000 years old, at Huatta, Peru.

RAISED FIELD TECHNOLOGY

KING'S SOAS STUDIES IN DEVELOPMENT GEOGRAPHY

Series Editors:
Robert W. Bradnock and Kathy Baker-Smith

Both the School of Oriental and African Studies and King's College, whose geography departments have recently merged, have established international reputations for their research into these areas. This series publishes original research into all aspects of geography in the developing world, particularly linking environmental and development issues. It will be of critical interest to geographers and academics in the fields of development studies, political science, environmental studies and economics.

Also in the series

Agriculture and Rural Development Planning
A Process in Transition
H. David Akroyd

Plotting, Squatting, Public Purpose and Politics
Land Market Development, Low Income Housing and Public Intervention in India
Robert-Jan Baken

Privatization and the Crisis of Agricultural Extension
The Case of Pakistan
Andrew P. Davidson and Munir Ahmad

Political Ecology and the Role of Water
Environment, Society and Economy in Northern Yemen
Gerhard Lichtenthäler

Planning in Contemporary Africa
The State, Town Planning and Society in Cameroon
Ambe J. Njoh

Participation, Negotiation and Poverty: Encountering the Power of Images
Designing Pro-Poor Development Programmes
Fletcher Tembo

African Environment and Development
Rhetoric, Programs, Realities
Edited by William G. Moseley and B. Ikubolajeh Logan

Raised Field Technology

The Raised Fields Projects around Lake Titicaca

ARTHUR MORRIS
University of Glasgow, UK

ASHGATE

Published by
Ashgate Publishing Limited
Gower House
Croft Road
Aldershot
Hants GU11 3HR
England

Ashgate Publishing Company
Suite 420
101 Cherry Street
Burlington, VT 05401-4405
USA

Ashgate website: http://www.ashgate.com

British Library Cataloguing in Publication Data
Morris, Arthur S. (Arthur Stephen), 1936-
Raised field technology : the raised fields projects around Lake Titicaca. - (King's SOAS studies in development geography)
1. Tillage - Lake Titicaca Region (Peru and Bolivia)
2. Sustainable agriculture - Lake Titicaca Region (Peru and Bolivia) 3. Agricultural development projects - Lake Titicaca Region (Peru and Bolivia) 4. Agricultural systems - Lake Titicaca Region (Peru and Bolivia) - History 5. Indians of South America - Agriculture - Lake Titicaca Region (Peru and Bolivia)
I. Title
631.5'8'098412

Library of Congress Cataloging-in-Publication Data
Morris, Arthur Stephen.
Raised field technology : the raised fields projects around Lake Titicaca / Arthur Morris.
p. cm. -- (King's SOAS studies in development geography)
Includes bibliographical references and index.
ISBN 0-7546-1790-4
1. Raised field agriculture--Titicaca Lake (Peru and Bolivia) I. Title. II. Series.

S494.5.R34M67 2004
631.5'8'098412--dc22

2003060977

ISBN 0 7546 1790 4

Printed and bound in Great Britain by Athenaeum Press Ltd., Gateshead, Tyne & Wear.

Contents

List of Figures *vi*
List of Tables *vii*
List of Illustrations *viii*
Acknowledgements *ix*

1 Introduction 1

2 Discovering the Past at Tiwanaku 4

3 Modern Research on the Raised Fields 15

4 The Projects for Reconstruction 37

5 The Situation Today: Some Case Studies 71

6 Agroecology and the Raised Fields 89

7 Some Final Lessons: Diversification and Integration of Farming 102

Bibliography *119*
Index *125*

List of Figures

3.1 The prehistoric raised fields area around Lake Titicaca. The estimates of Denevan, based closely on aerial photography, are much more modest than those of Kolata. 17

3.2 The daily temperature range is a major restriction on crop cultivation in the mountains of the tropics. Even near the lake it is a stress factor for plants. 21

4.1 This shows the early project areas up to about 1990. They cluster heavily in the core area around Huatta. 43

4.2 The prehistoric raised fields area around Lake Titicaca. The estimates of Denevan, based closely on aerial photography, are much more modest than those of Kolata. 44

4.3 The early plans of PROSUKO were large scale and involved the whole Bolivian shore of the lake. 51

4.4 By 1994 the areas had been refined and reduced, but some priority areas never had projects on them. 52

4.5 The construction of raised fields. A diagram used in the field to help farmers shows dimensions in metres, and materials, including the reinforcement of field edges with cut turfs. 56

List of Tables

3.1 Juliaca – rainfall and crop requirements (measured as potential evapotranspiration through the growing season) 23

4.1 Value of family agricultural (non-livestock) production in 1995 & 1999 (in constant soles 1995) 64

List of Illustrations

1.	Remains of ancient raised fields, possibly 2000 years old, at Huatta, Peru.	Frontispiece
2.	A raised field with bitter potatoes, in a project community run by PIWA, at Pomata. The PIWA agent, Ernesto Chura, accompanied and helped the author in the field.	38
3.	The author at one of the earliest raised fields projects near Huatta, now abandoned.	38
4.	Harvest time at Huatta, near Puno. Barley has been cut by hand and laid out to dry before putting into stooks. All harvest work is by hand and in this system, the reintroduction of an ancient technology might seem a godsend, allowing more production without recourse to modern machinery.	39
5.	Shepherd girl and sisters at Chucuito, south of Puno, watching a lamb in the totora reed beds by Lake Titicaca, while their father is out fishing. Livestock and fishing are important elements in the farming system. Diversity and using alternative resources are fundamental to these people's economic strategy.	114
6.	Lima Pampa, a rural settlement north of Juliaca, Peru. On this treeless plain there are no good fuel sources and the animal dung is dried on the adobe walls around the village huts, to be burnt in open fires or stoves for cooking. Farming here is more geared to livestock than arable, which might have had a better use of the dung for manure.	115

Acknowledgements

My first debt, for making me aware of the remarkable achievements in restoration of raised fields, is to Oswaldo Rivera, head of the National Department of Archaeology in Bolivia, who lectured on his work of raised field restoration in Bolivia at a conference in Finland organized by Sakari Tuhkanen, and inspired me with a little of his missionary zeal for the subject.

Considerable early support was given by Chris Brown, representative of the British Council in Peru, followed by Eric Lawrie, as head of the new office set up by the British Council in Bolivia, which fostered a formal link between my University Department and several Bolivian institutions, over 1998-2000. The Metropolitan University of San Andres, UMSA, in La Paz, and its head of geography, Ingeniero Reynaldo Oroz, jointly with the School of Military Engineering, or EMI, especially with Colonel Alberto Abastoflor and General Alfaro, were involved in this link, as well as the Catholic University.

Following the support from the British Council, my research work has been financed by the Leverhulme Trust who gave me a Research Fellowship, and then an Emeritus Research Fellowship when I revealed that further work would be needed because of national strike in Bolivia on one visit, and a catastrophic flood period in the altiplano on another, both of which made field work virtually impossible. I am indebted to them for their support.

In the field areas, I had special help from the research director of CARE International at Puno, Peru, Alipio Canahua, who helped me to understand the large and then successful programme of raised field restoration undertaken by his team, and indicated some research directions. Further important help, including access to their considerable library, came from the agency PIWA in Puno, and I am indebted to its head, Dr. Hugo Rodriguez. The publications from this institution remain the most important sources of information on the whole programme of restoration. Dr. Edgar Avila, Vice-Rector at the University of the Altiplano in Puno, and then one of his staff, Edgar Gomez, helped through organizing a group of students into studies of several village communities.

In Bolivia, the greatest support was from Colonel Abastoflor, and then from Eduardo Chilon, university lecturer at both the Catholic University and EMI in La Paz. Eduardo continues to investigate the subject and to cover the large and poorly understood field of Andean technology. He was able both to direct me to field sites of interest and to publications on the subject, including a variety of dissertations by his students.

My academic colleagues were always supportive of my work, especially those who came to visit and do research and lectures at La Paz, David Forrest, Stella Lowder, Jane Drummond, and Jane Benton. At the University of Glasgow, I owe special thanks to Karen Hegyi, who prepared the document for publication very efficiently, and to Mike Shand, who drew the final versions of the maps.

Apart from all these, there were many unnamed helpers in Bolivia and Peru, and in Britain, who either joined in my enthusiasm or simply kept me going over what has turned out to be a long journey of discovery. Some of them may read this book and will have to forgive my failure to add their names to this note.

Arthur Morris

1 Introduction

This book is an evaluation of a great experiment, an attempt to bring back a lost technology, for a region whose rural inhabitants are some of the poorest on the South American continent. Starting in the late 1960s, and carrying on through the 1980s, scientific research resulted in the rediscovery of a lost technology, the raising of the soil into long ridged cultivation platforms, around Lake Titicaca, and led to major efforts at the reintroduction of this technology to the farming lands of the high plateau around the lake. At an experimental level, yields of the main crop, potatoes, and of several other important local crops, were found to be more than double the local norm, because creation of raised fields improved either soil structure, soil drainage and flood control, temperature control, or all of these. Great excitement was created by the findings, and plans were laid for programmes of rebuilding and reuse of raised fields in rural communities, typically on collectively owned lands where all the able adults of the community would work together, as was thought they had done in prehistoric times.

There are no other areas in the world with climatic conditions comparable to the high plains around the lake, and so the new found technology cannot be applied baldly to anywhere else. This review of the efforts over the last two decades does however have relevance to a variety of rural farm development projects, which seek to use indigenous knowledge, or to provide sustainable development for people now in danger of being overwhelmed by modern farming and its dependence on outside knowledge and outside capital resources.

The raised fields, termed *waru waru* in Quechua, the language of most of the Peruvian area of concern, and *suka kollus* in Aymara, the dominant language in southernmost Peruvian sector and into northern Bolivia, were first identified from their eroded traces in the ground around Lake Titicaca, as well as elsewhere in Latin America, and these traces can be identified from aerial photography covering 82,000 hectares. Because of later cultivation, and recently the expansion of the main cities, some have been obliterated and their original extent was probably much greater, over 100,000 hectares. Today, they are slight undulations in the soil, no more apparent than the old plough ridges of Southern England's medieval three field system, or the lazy beds used by the crofters in nineteenth century Scotland.

But originally, they were probably up to a metre in height, with the intervening ditches often water-filled during the growing season. These were cultivation platforms, three to five metres wide, 20 to 100 metres in length, set in small groups or bunches of five to 20 at a time, oriented in many different directions. They covered huge areas of the lakeside plains, a massive bewildering patchwork, extending even below the present level of Lake Titicaca in places. Their purpose was to check night time frost on the plateau, frost which can occur at any time of year and damage tender growing crop tips. They also helped soil quality by deepening and aerating the top soil, providing also more warmth and better soil drainage. In some cases they were a guard against floods. Though they had mostly been abandoned for 800 years or more, these raised fields held huge promise for a gigantic experiment in reconstruction which might be held out as a solution to the trap of rural poverty leading to low production which in turn allowed no improvements or modernization of farming in the region.

This project is obviously worthy of evaluation, because it would bring back a lost technology, one of great promise, but also because it would use locally known ways of working, which had at one time provided a sustainable livelihood for a large population, and could support a sustainable development in the future. But the intention is not a standard evaluation such as is done regularly by agencies bringing in new development projects. Such an evaluation might be in economic terms, showing the physical increase in crop yields, the surplus of benefits over costs, the changes in farmer welfare over the period of a project. Nor does this work claim to be the somewhat broader social-economic type of analysis. This would have to include the changes in family organization and work patterns, membership of groups and organizations, linkages with agencies that have promoted the technology, alongside the purely economic measures. All these kinds of studies have been done and indeed are reviewed in the text, but they cannot give a full picture of the sustainability and value of the technology.

Instead, after several successive visits and substantial study of the agencies and the target communities involved in the experiment, I have come to the view that the most useful review is one which examines the history and prehistory of the technology, the way it was brought back into public attention, and then looks at the actual programmes and relations between agency and community, and the way the raised fields were actually incorporated into the farming system. Many anthropologists and sociologists have commented on the holistic approach taken to farming by the native communities in the Quechua and Aymara speaking world of the high Andean plains. Given this holism, it is necessary, I believe, to establish how the projects fit into the farming and life systems of the people, to try to see whether they have been incorporated in any real way into the yearly cycle, and into the collective mentality of these people.

I have used the valuable research documents of the main agencies, especially those of PIWA, the agency which managed the projects for the government in Peru, many of which were published as books over the period of 12 years from 1990 to 2002. In addition I had the advantage of being able to travel out and visit a large number of the communities affected by projects, with agents and ex-agents of the programmes, as well as with academics from the local universities. Apart from this, my information was gleaned from a number of visits made privately, to talk with community leaders, and with a variety of farmers who had been involved in the projects. From this I could learn of the very different viewpoints about the projects in the minds of experts from outside, local agents and agencies, and the farmers themselves.

The major programmes of reintroduction are coming to a halt after about 15 years of effort, so that some kind of evaluation is now possible. One conclusion, however, will be that because of the long cycles of land use typical in the region around the lake, a final evaluation is not possible and will only emerge after many years. In this region there are ancient terraces, used for a crop or two and then left, apparently abandoned for twenty years or more. The evidence of these high terraces shows us that success or failure is not to be measured over time scales of four years or so, the typical length of an agricultural project. A related conclusion will be that what looks like total failure and abandonment in many areas, after a few years of projects, may in fact not be a failure but a filing in the collective memory for use on occasion and when circumstances dictate. The local evaluation can differ greatly from that of outsiders.

2 Discovering the Past at Tiwanaku

The tourist who arrives at La Paz airport, appropriately called El Alto (The High), at 4000 metres above sea level, has no difficulty in comprehending the poverty of the high plateaus and mountains of Bolivia and Peru. Apart from a numbing coldness of the air (most international flights from the Northern Hemisphere arrive in the early morning) and nausea from the altitude, the area which is visible around El Alto gives the impression of a barren, unproductive land, treeless, flat and either dusty dry or so wet that water lies everywhere in muddy pools. The trip down to the city through shabby self-build suburbs of apparently half built little houses, does nothing to cast doubts on the first impression of poverty. Yet first appearances can be deceptive. At one time, the high plateau or altiplano was the centre of relative wealth, and of an ordered civilization, controlling a large part of the Central Andes. Tiwanaku[1] is the name of the ancient centre of culture, and this name is used both for a small town created in colonial time, now standing a few miles from Lake Titicaca, and, less than a mile away, for the remains of a great city. This city had palaces, some erected on the flat tops of pyramids of earth, possibly faced with stone and with terraced sides. It had semi-sunken courtyards reminiscent of the Maya cities, and evidence of water management systems including canals and piping for water drainage. Around the ceremonial centre of the city there were large areas of mud huts occupied by the majority of the population.

Over the past half century, the remains have been sifted over by archaeologists and some questions about its little known culture have been resolved, including a chronology for its rise and sudden downfall, and some elements of its material culture. There remain many mysteries about Tiwanaku however. The building techniques, which were adopted in part by the later Inca civilization, are imperfectly known, especially the bringing of giant stones over long distances to the site, and the stone cutting and

1 The spelling of Tiwanaku is the most common accepted form today. However there are many sources which still use Tiahuanaco, Tiahuanacu, or other versions.

fitting together of blocks with fine tolerances so that they effectively form dry stone walls, but with no gaps. A central query is how the considerable population of this state or proto-state of Tiwanaku could have maintained itself in food and clothing for some 800 years, given the apparent hostility of the environment. To maintain its urban centres – the city of Tiwanaku itself may have had 150,000 people – required a highly productive agriculture, producing surpluses to support a caste of non farmers, the priests, kings, and their entourages, as well as many artisans for building, pottery working and the like. This question is addressed in the present book. A fascinating idea is that this culture had a successful agricultural technology that could be brought back with great benefits for the modern inhabitants. An intensive high yielding farm system which could support a large non-agricultural population is certainly something better than the present state of affairs, in which small farmers produce only enough to satisfy their own needs for sustenance, plus an occasional meagre surplus for market. Today, many farmers have to move out to the coastal region where they can find paid labour in the sugar and cotton harvests, and come back home after six months with a little money.

The Discovery of Tiwanaku and its Culture

Scientific study of Bolivia's prehistory dates largely from the twentieth century. During the colonial period, Spain occupied all western South America, and ruled what is now called Bolivia, then the region of Alto Peru or Upper Peru, as a resource frontier. The Spanish exploited the silver mines, especially those of Potosí – a mountain filled with silver veins in the southern altiplano – sending large amounts of silver back to Spain, but maintaining the population in a condition of virtual serfdom, whereby the rural communities were forced to supply men to the mines where they suffered cruelly through the harsh working conditions, or keeping them as serfs on the great estates, the *haciendas* which produced grains and wool for trade. The natural environment, and particularly the climate, of the altiplano, was foreign to the Spaniards; no European region has cultivated land above 2000 metres, let alone the 4000 metres of the altiplano.

Specifically, European incomers evidently had no understanding of technologies like those of the raised fields, which are perhaps most effective in this specific environment and nowhere else. During the colonial period, it is likely that some small areas of raised fields were still visible and in use, cultivated by peasant farmers outside the great hacienda system. But most would have been destroyed and lost, even by the Incas who were the imperial power in this region prior to the Spanish. Inca culture was centred far to the north, at Cuzco in the Peruvian mountains, where the climate is less rigorous and more humid and where the raised fields might have had limited value. In any case, neither Inca nor Spaniard is likely to have taken up the practices of a conquered people, when they did not understand them or see the need for them. Some Peruvians believe that the Incas also understood and used the raised fields extensively (Espinoza Soriano 1997), but there is no proof of this. Neither Inca nor Spanish colonial sources comment on the use of these fields, which is remarkable if they had remained in important use.

As to the main urban remains of the culture, Tiwanaku was known about, but did not attract much interest, in great part because the stone monuments were half encased in windblown and stream deposited sediments. Up to the 1960s, they looked like a set of monoliths on the open plain. What lay below was, in reality, a palace complex with pyramids, sunken courtyards, canals for water provision and drainage channels, with a substantial city of mud huts surrounding it. Famous travellers such as Cieza de Leon and Bernabe Cobo made useful observations on the apparent great age of the ruins, and the latter, exceptionally among early writers, thought that they actually represented a great city (Albarracin-Jordan 1999). But the later colonial period saw a complete hiatus in study, possibly due to the state of political turmoil and constant rebellions in the Central Andes at that time. In any case, any attempts at study during the colonial period were few and far between, and limited in scope. The early *conquistadores* had looted the gold, silver and jewels of Inca civilization in Cuzco and elsewhere, and had little regard for anything which could not promise rapid wealth. Traces of ancient farming structures would not have interested them in any way. Their successors, the landowners of the colonial period, were often absentee owners, the merchants or administrators of colonial cities, and were not involved directly in farming the land or even supervising the farming, which was left to overseers.

In the nineteenth century, the situation changed little; after Independence in 1825, Spanish influence was gradually replaced by that of northwest Europe, and the silver mines eventually gave way to the tin mining whose acme was reached in the early twentieth century, along with mining of other ores. Little attention was paid to local farming, let alone the interpretation of a forgotten farming system, which might only have benefited the rural poor. During the nineteenth century there were indeed visitors, but their interpretations of the Tiwanaku area and its environment were negative. George Squier (1877) typifies the efforts of this century. He was from the United States, appointed as diplomat to Lima in 1863, with strong archaeological interests and preceding publications on the archaeology of the Mississippi river valley. His view, briefly, was that Tiwanaku, then merely an extensive area of giant stones half buried in silt blown in over centuries, was a largely ceremonial site, with no large permanent population. He was impressed by the current poverty of the altiplano, the lack of arable farming, the harsh climate, and the evident backwardness of the inhabitants of the region. By the early twentieth century, the view on this site had expanded to allow that perhaps Tiwanaku was a true urban centre, though without fully recognizing it as the centre of an important civilization. Max Uhle (Stubel and Uhle 1892), perhaps the most respected archaeologist to visit the site before the 1930s, conceded an urban form, but did not appreciate the size and importance of the civilization which lay behind this city. These writers also thought that the Tiwanaku civilization was the product of the Aymara-speaking people, which is a significant contribution. Later, work on Tiwanaku and associated features such as the raised fields was to be associated with Bolivian nationalism, or rather with a nativist force, seeking the recovery of the Aymara culture as central to Bolivian national identity.

Arthur Posnansky

First twentieth century interest in Tiwanaku came not from traders or farmers or the governments of the day, nor even from formal science, but from an amateur, an adventurer, who stimulated local interest through his exuberant enthusiasm for the past. Most of Arthur Posnansky's ideas about Tiwanaku have been proven wrong, but his central idea, that here there had been an important civilization with great technical skills and accomplishments is now proved correct. We can trace his history largely from the intellectual biography written by Ponce Sangines (1994). Posnansky was born in Vienna, the son of a wealthy industrialist After studying naval architecture at Pula, on the Adriatic, he set off for the tropics at the height of the rubber boom, in the 1890s. Having established himself in trade, he bought an 18 metre launch in 1898, for freighting supplies upstream to the rubber lands, and rubber downstream, to the main centre for the rubber trade, Manaus. At this time, there was a rapid advance up the rivers to find the sources of wild rubber, and the Brazilian frontiersmen were moving up the Acre river into what was Bolivian territory, though Bolivia was unable to control this distant corner of the republic. The Brazilian incomers wanted independence from Bolivia from 1899, and some fighting broke out. In 1900, Posnansky offered help to the Bolivian cause in maintaining control of the area, was captured by rebels and escaped. At Bolivian requests, he went so far as to take his vessel back to Germany to be converted to a military vessel to help keep the peace in the rubber lands of Acre, but found first, that his vessel was confiscated by the Brazilians in Manaus, and then that the Acre territory had been sold, first to a private company, then passing to the Brazilian state.

Posnansky eventually moved to La Paz in 1903, to reclaim his investments in the boat, some £17,000, and after long negotiations was awarded £11,460. His stay in the city was long enough to get him interested in establishing himself permanently in Bolivia. La Paz was only a town of 52,697 inhabitants in that year, and Posnansky made a major impact by joining and becoming prominent in the nation's Geographical Society which had been founded in 1899. This society included in its members much of polite society and politically important people of La Paz. One of his topics for addressing the Society was Tiwanaku. He had first visited it through a request by the then President of Bolivia, President Pando, in 1903, for Posnansky, a noted shipbuilder, to assemble and launch a motor boat at Guaqui, Bolivia's new port on Lake Titicaca. On his way to the port, using the new railway which connected it with La Paz, Posnansky passed through Tiwanaku and visited the ruins. This chance event set him on an intellectual career which was to occupy the rest of his life. Tiwanaku was already being studied by a French expedition at the time, but this group had little scientific discipline and was actually destroying some of the evidence, and doing little to excavate the ruins, still more than half buried with silt. The railway to the lake, built over 1900-1903, had already destroyed much by ploughing a line through the edge of the site, and the colonial authorities had earlier used many great monumental stones to build the church and other buildings of the town of Tiwanaku.

Posnansky lost little time in publishing ideas about the origins of Tiwanaku, but also demonstrated one of his fundamental failings, a willingness to accept folk legend as scientific truth. His first contribution to the Geographical Society was based on the idea that the stone work was the result of molten lava from a volcano in southern Peru, poured into prepared moulds – an impossibility technically, but also geologically because the volcano in question had been dormant for millions of years. Almost equally unacceptable were ideas he developed for a conference in 1908, when he claimed Tiwanaku had been a city on an island in Lake Titicaca, and for the important International Conference of Americanists in Buenos Aires of 1910, where his talk included the idea that Tiwanaku had benefited from a much more benign climate in its early years, because the cordillera of the Andes had not yet risen (Ponce Sangines p.32). Again, the ideas were geologically impossible, because of the time frame he proposed. Another concept developed at this time was that Tiwanaku was incomplete because an earthquake had destroyed it and let in the lake waters. Not that Posnansky was alone in wild theorizing; in 1911, the Norwegian geologist Laurence Sundt had claimed that the great blocks of Tiwanaku, some of which were brought over many kilometres, were brought to the site by icebergs in the glacial period.

Eventually Posnansky brought a German wife out to La Paz, established himself with a steady and profitable firm, making bricks and ceramics, and was thus able to finance his intellectual interests which were encyclopaedic but centred on Tiwanaku. He achieved an important position in La Paz society, as President of the Geographical Society, organizer of the first films in La Paz, the first person to bring a car to La Paz, and the creator of what would become the National Museum of Archaeology in La Paz. In 1937, he was also the first to suggest a national park in Bolivia, to include Tiwanaku. It would have been peopled with what he considered to be pure Indian races – the Uru and the Kollanas – and to be forested with native species, the *queñua* (Polylepis) and the *kiswara* (Buddleia), and have guanacos and vicuñas running wild in it. This must surely have been the only national park suggestion to include human specimens as part of the attractions! In the event, Bolivia's national parks have been established much later and in the eastern forests, and the altiplano remains seriously lacking in conservation measures of any kind.

Some of his ideas were more dangerous. Posnansky was evidently in touch with current German ideology and in typical fashion, seems to have swallowed it whole. He saw an American equivalent for the *Herrenvolk* of Europe, in the Kolla race (never really identified by any reputable anthropologist as a separate physical race) a non-mongoloid, broad headed group, to be found throughout the continent, and a race which he posited as native, i.e. not coming from Asia as is the standard theory, but emerging here in a heartland of early mankind. This leader race, as he developed the idea in his last years, very much aligned with Nazi ideology, became the builders of Tiwanaku, and their culture would permeate all of early America. This kind of thinking and writing were accepted by some Bolivians at the time, firstly because Bolivia sided with Germany though a neutral in the war and there was an influential German community in La Paz. It also suited Bolivian nationalism to find their country as central to civilization, in contrast to the cultural dependency on Spain inherited from colonial times.

Overall though, Posnansky's ideas were always contentious and many were challenged during his lifetime, to be finally dismissed over the following decades, but his importance as a pioneer is enormous. Most centrally, he set up an impetus for the study of this pre-Inca culture which has continued to the present day, and which included the task of finding how such a civilization could find physical support in the high plateau country of the lake. An archaeological society was established in La Paz under his direction, and the university introduced the subject for students through his intervention.

Contributions from Others

Systematic archaeological digs first took place at Tiwanaku in 1932, under Wendell Clark Bennett (Bennett 1934, 1935). This serious, scientific work established a chronological sequence for ceramic remains, with the earliest period starting about 500 BC, and the culture coming to an end about 1200 AD. Bennett also recognized that this had been the city central to a large civilization, spread widely through the Andes over a substantial time period. From Bennett's time forward, and especially after the National Revolution which overturned traditional authority (the church, the landed gentry) in 1952, there was a growing feeling of nationalism and a need to find symbols for this. One of these was to be found in Tiwanaku and the Aymara. Bolivian archaeologists were formed and began work on the subject, institutionalized in the *Centro de Investigaciones Arqueologicas en Tiwanaku* (CIAT), from 1958, with governmental support. The leader of this work, Carlos Ponce Sangines, extended Bennett's work by establishing an early settlement phase reaching back to 1600 BC, and by discovering a phase of urbanization, an "urban revolution", which made a real city of Tiwanaku, about 0-100 AD. Sangines also conducted digs in other sites which proved the existence of an imperial phase from the eighth century AD to the end in 1200 AD.

Posnansky had never thought of Tiwanaku as more than a ceremonial or religious centre, with little in the way of permanent population. Sangines corrected this view. His estimates of the size of Tiwanaku's population underwent successive increases as more was unearthed around the central ceremonial site and a city of up to 150,000 is now accepted as likely. In the Tiwanaku empire, central influence extended over an area of 600,000 sq. km. and a population of 3.6 million (Albarracin-Jordan 1999, p. 25). The size and urban complexity of the empire are shown up in the work done outside Tiwanaku, such as that of Alan Kolata in the plains of Pampa Koani, a few miles east of Tiwanaku. Here, a number of second order urban centres, with temples and administrative buildings, were discovered to be of the same age as Tiwanaku, and roads and canal systems connected with the use of the farmlands and linking them to the towns is also known. Evidently, the great central city was nourished, as elsewhere in the world, by a network of other towns and by a fertile agricultural hinterland. Our trail leads inevitably to the need to define the food support of this important urban population.

False Trails

Kolata's work will be reviewed in the next chapter. But before his time, interest in the countryside was considerably less than in urban development, and negative in its conclusions. In the 1950s, a noted French anthropologist, Louis Baudin (1955, 1960) had regarded intensive agriculture as something impossible on the altiplano, and found it hard to explain just how great civilizations could have come into existence at all in this environment. In his words:

> It is difficult to imagine that this region could have been the centre of a great vanished civilisation. The climate here is so harsh that maize does not ripen; the vegetation is so poor that only the wild olive* grows to shrub size and there are no trees; the lake weather is so cold that man cannot bathe in it without danger. Torrential rains, burning sun, freezing nights follow each other in sequence, even in the presence of a water mass which diminishes temperature variations and makes the climate of the lakeside fringe more temperate than the puna.
>
> * this must refer to either the *Buddleia incana*, which is called *olivillo* in South America, or *Polylepis incana*, the *queñua.*

There were also anti-scientific views about the Tiwanaku people, which certainly did not help the real work being done. The remaining mysteries about the origins of some features of Tiwanaku, such as the movement of the great stones and their fashioning into palace walls and temples, have allowed the flourishing of a science fiction genre which has obscured scientific progress. Thus for example, Bellamy and Allen (1948) published a book attributing the building of the monuments to extraterrestrial beings, and Eric von Daniken (1969) used the same kind of thinking, at a time when science fiction was particularly popular. Not that this kind of publication has ceased. A recent contribution is J.M. Allen's (1998) book suggesting Tiwanaku as the oldest archaeological site on earth and the birthplace of mankind, using this site as the origin of the Atlantis story in Greek legend. A more plausible, but ultimately teleological and scientifically objectionable set of arguments about this site comes from the work of Maurice Cotterell (2001). He tries to link the Tiwanaku culture and its special technology of construction, through the God-King Viracocha, with that of Sipan, a series of mud block pyramids only recently excavated by archaeologists, on the north coast of Peru near Chiclayo, and the mysterious lines of Nazca traced in the desert on the southern coast. There are certainly obvious links in the art forms and symbols used throughout the Americas, but this is scarcely a mystery given the recent migration of most native Americans from the Old World, in the last 20,000 years. These people are likely to have maintained certain elements of material culture in common, simply because of their common origins. Sipan, Nazca, and Tiwanaku have many unsolved problems about their meaning in American prehistory, but none which require any recourse to extraterrestrial forces.

The Peruvian Alternative

There has been a rapid evolution of thought about the Tiwanaku, mostly as a result of Bolivian archaeology and its rapid rise in recent years. From a set of interesting ruins, the twentieth century has identified this as a major civilization, probably inspiring and providing the antecedents for the Inca empire which seems to have taken over its territories and its achievements, in the interval from 1200 to 1450 AD. Inca invasion and total replacement of Tiwanaku power did mean that later observers have found it hard to trace the technology and distinctive achievements of the Tiwanaku before them.

Tiwanaku was of course not an isolated city, or an isolated civilization, on the altiplano. There are other traces in the sand, parallel to this culture, in other locations, or at other time points. It is of interest to point out just one of these, as it may have been equally important in the development of farming in this region. To the north in Peru, and still close to the lake and our main area of interest, there was the culture of Pukara, which as a high culture was antecedent to that of Tiwanaku in this area and which was also associated with agricultural development, including raised fields and *qochas*, ponds used for agriculture.

Thus, while not challenging the importance of Tiwanaku overall, some writers have maintained that in the matter of agricultural technology, other cultures may have been earlier and more significant. Berastain (1999) believes that the maximum period of raised field construction in the northern area around Pukara was between 800 and 600 BC, in other words hundreds of years before the rise of Tiwanaku to urban status. The central ruined palaces and temples lie next to the village of Pukara, some 30 km north of Juliaca and Lake Titicaca. The archaeological site is above the present village, on a moderate slope rising above the Titicaca plain and sheltered by a high cliff behind. There is a small and rustic museum, with a single keeper, and the ruins, smaller and far less impressive than Tiwanaku, are virtually unguarded. Unlike Tiwanaku, this site is unknown to the average tourist. But it is certain that this culture was associated with remains of high ceramic art and skill in engineering and architecture. Given the early date of development of this centre, and the presence of old raised fields, it is likely that this area had a close association with the early spread, if not the origins, of this farming technology.

For Erickson – the archaeologist who has done most work on the Peruvian raised fields – the origins are however not in any central power and its imposition of a farming technology, but in small scale farming by the lake, which was then extended and developed by the Pukara culture as part of the Peruvian Early Horizon (1200-200 BC). Pukara reached its maximum cultural expression between 200 BC and 550 AD, and possibly achieved the level of organization of an early state (Berastain 1999, p. 25). There are nationalistic overtones in the differing views of development of agriculture between Bolivian and Peruvian versions. Both sides would want to claim the early development of the raised fields but there is also a technological significance in the opposing claims. If the Peruvian claims of Erickson are correct, then the raised fields develop within a village culture and without the need for state management. If the Bolivians are to be believed, then the great state of Tiwanaku and its potential for mass labour organization was needed before the raised fields technology could assume any importance. We shall develop this theme in the next chapter.

There are also raised fields in various other parts of South America, and indeed in North America, discovered over the period from the late 1960s to the 1980s, which begs the question of whether this was a practice brought from elsewhere by migrants, or expanding empires, and under what conditions the introduction took place. We can only speculate on this topic, since there are no conclusive direct links from these other centres to the altiplano farming. It is true also that the uses of raised fields were different in the other areas, such as the northern plains of Colombia, the lowland plains of the Guayas, the eastern plains of Mojos in Bolivia itself, and elsewhere (Denevan 2001). In the lowland tropics, drainage was probably a key function of raised fields, whereas temperature control was vital in the highlands.

In effect however, what was most important was the way in which the techniques of construction and use developed in the altiplano itself, as this historical path may suggest more about the possibility of reviving the practices today.

3 Modern Research on the Raised Fields

Introduction

Since the 1960s, an enormous amount of work has been done in reconstructing the history of the Tiwanaku civilization, and with it, the special technology which seems to be associated with this epoch. The chapter reviews the experimental archaeology of rebuilding raised fields, and the agricultural results of crop trials using raised fields. There were also detailed climatic and other physical studies done, to indicate how the technology may have allowed exceptionally high crops. We also examine the debates between different scientists, especially between the major research groups led by Kolata and Erickson, on the Bolivian and Peruvian sides of the lake respectively, about how the technology began and how it ended in antiquity. These debates are important for the current efforts to reintroduce the technology, because they may help identify factors of success and failure which are as valid today as they were in the prehistoric past.

The Twentieth Century Before the 1960s

There was no scientific research on the raised fields phenomenon until relatively recently. Those archaeologists studying the Tiwanaku who were most closely associated with the fields were always more concerned with buildings and settlements, and the significance of the traces of ancient field forms escaped virtually all other commentators. In 1923, Max Uhle, a well trained German archaeologist resident in Peru who had visited Bolivia in 1910 and had crossed swords with Posnansky over some of his more extreme claims, had noted the existence of the raised fields, but had not conducted any field research into them. However, in the second half of the twentieth century, a body of research would emerge which would reveal a technology spread widely in the Americas, and possibly of central importance to the interpretation of some ancient cultures. It would also inspire the efforts made to reintroduce the raised fields as a programme to help farmers in Bolivia and Peru. In this chapter we review this history of research, with special attention to the ideas which grew up about the association between the kind of state which may have existed and the organization of land management. Specifically, if a strong, despotic state was needed to construct fields, then perhaps even today a major centrally organized effort would be needed by the modern states of Bolivia and Peru, to re-introduce the technology. Alternatively, a more pessimistic view is that any modern introduction would be impossible, as the social systems of the despotic state (slavery, serfdom?) were not now possible to bring back or are acceptable.

The first studies on the raised fields were to produce a revolutionary rethink on the farming of the altiplano. In the last chapter, we noted that as late as the mid-1950s Baudin had regarded this region as incapable of sustaining continuous cultivation. In the 1960s, the idea would come into circulation that there was continuous and intensive cultivation of the plains around the lake. First scientific reports of raised fields came not from Titicaca but from a still more remote region, the savanna grassland plains of Mojos of eastern Bolivia, along the southbank tributaries of the Madeira, the Mamore and Guapore; here they were studied by George Plafker (1963) and then by William Denevan (1966); they were correctly identified as agricultural engineering structures from the period before European discovery, and linked to the need for dealing with soils in a region of flat plains which is seasonally inundated. Raising beds here puts them above the floods and allows cropping. Mucking the canals out onto the ridges improves the soil on the latter, and possibly allowed permanent cultivation of manioc, the main starch provider and alternative to grains in the tropical lowlands. As proved the case elsewhere, there was no single continuous orientation or single pattern, and the orientation is never directly downslope, so that drainage cannot ever have been the sole purpose.

Following this discovery, raised fields of comparable type were reported on from a variety of floodplain sites, in the Guayas valley of Ecuador's Pacific coast, and in the plains of Mompos, northern Colombia, in both of which sites there is also seasonal flooding. These and other lowland areas are reviewed by Denevan (2001), showing the upsurge of interest in the phenomenon in the 1960s and 1970s. For the Titicaca region, the first published study is by Smith C. *et al.* (1968), who mapped the area of remains from aerial photography and field reconnaissance. They describe in detail the total area involved, and note the major concentrations, the area around Huatta to northeast of Puno, and the smaller areas around Pomata and the Desaguadero all in Peru. On the Bolivian side, they are clustered north of Tiwanaku, and near Aygachi (Figure. 3.1). All the areas are on flat land, where drainage and flood control would be of relevance, and within a few miles of the lake, where the most moderate climates are found.

Figure 3.1 The prehistoric raised fields area around Lake Titicaca. The estimates of Denevan, based closely on aerial photography, are much more modest than those of Kolata.

C. Smith also describes gaps around the lake on the west side where the most suitable flat land is found, but where there may have been raised fields in antiquity that have been erased later. What seems to have happened is, after abandonment of the raised fields, their replacement by great estates, right through the colonial period and up to modern days. This kept the trace of the raised fields on the old estate lands under a protective and permanent grass cover. By contrast, in some areas such as around Ilave and Taraco, there was a dense indigenous population and cultivation was intense, erasing most remains of the fields.

C. Smith and his associates noted the wide diversity of field types and tried to classify them. The most important type was what could be called an open chess board, with bundles of five or six ridges in one direction, at right angles to the next bundle and in a loose, open arrangement, with water channels linking through all the bundles. Another common type is the irregular embanked type, with a bund or bank surrounding individual sets of five to 20 ridges, representing a family or small community's attempts at flood prevention. The purpose of the fields was sometimes apparent, (flood control in the last type) but was usually difficult to decipher. The reclamation of marsh land was seen as important, irrigation difficult to see, and thermal considerations thought to be secondary.

Other writers and researchers looked more closely at the raised fields in the highlands of Lake Titicaca and elsewhere to find a key to their purpose. Morlon (1981) regarded irrigation and drainage as the main element, and Lennon (1983) who was involved in field work in Peru, reckoned water control as the principal aim. R. Smith (1987) in a review of Andean technology, saw drainage and soil improvement as the most significant functions. Evidently, the purpose was not apparent or obvious to European and North American workers. For Northern Hemisphere scientists, irrigation and water drainage were familiar, but these South American structures did not conform readily to any known pattern.

Unlocking the Secret: Micro-climatic and Other Physical Effects

One way to work out the purpose of the fields, a method becoming accepted by archaeologists in the 1980s, was to attempt a reconstruction of them, and this was the next phase. Workers in various places including the two main groups of researchers, under American archaeologists Kolata and

Erickson, were able to do this, finding that they were successful in producing big crops, and identifying some of the effects they had on climate and soils. Before them, in Ecuador, Ryder and Knapp (1983) were the first to report a considerable thermal protection, in terms of air temperatures,[1] from building experimental raised fields in an area near Quito, where at 2850 metres, night time frosts may occur during the growing season. This effect occurs, first because of the high altitude and thin dry air, allowing strong outward radiation. And secondly, because wind speeds in the Central Andes are usually very low, notably at night time, and the air, chilled by contact with the cold ground, slips like water from the growing platforms into the ditches beside them.

In Bolivia, Ortloff (1989) made detailed studies, including soil temperature and air temperature as well as the temperature of the water in the ditches. This water, in season, might have temperatures as much as six to nine degrees higher than the adjacent air on cold nights. The emphasis given by Ortloff is however on the retention of heat by the soil in the raised fields, which protects the delicate roots of the growing crops from damage (Kolata and Ortloff 1989). Ortloff also commented on the greater sun incidence on the sloping edges of the raised fields, and the improvement of air content in the soil and soil structure through ridging. Many of the experimental raised fields, including those of Ortloff, were constructed under the big programmes of Alan Kolata, in the Pampa Koani, north of Tiwanaku, and in Peru, under Erickson, in the area around Huatta. These are good and relevant locations for such experimentation, in the sense that they were located in major areas of raised fields in prehistory, and enjoy climates and soil conditions which are typical of the circumlacustrine region.

The climatic and soil improvement effects of raised fields have been confirmed by other later experiments. Occasional contradictory results, such as that by PROSUKO, the Bolivian programme agency, showing the same air temperatures at night over both raised fields and open pampa, were shown to be due to placing the thermometers lower over the raised fields than over the pampa; the coldest air is nearest the soil (COSUDE 2000). This exceptional result was against a set of others which showed the raised fields to have a two degree advantage over the pampa.

1 All temperatures are in Centigrade unless otherwise noted.

This was from small experimental plots of about one to two hectares, some without standing water in the ditches. Even better results could be expected from larger concentrations of the fields and from ensuring water in the ditches at times of frost danger.

On the Peruvian side, the temperature differences between raised field and pampa are somewhat lesser, reflecting perhaps a climate which is slightly less extreme, more humid, and thus less affected by radiation cooling. At Caritamaya near Puno, the differences were from 0.67 to 1.36 degrees, in tests of 1992 (PIWA 1994). For the crops which are occasionally tried on the pampa, the bitter potato and the *kañiwa* crop, some degree of frost resistance is built in. A frost of -7 degrees Centigrade (a severe frost) will cause 80-90 per cent damage in normal potatoes. In bitter potatoes of the Ruckii species, the damage caused is only 10 per cent (PIWA 1994). Similar results can be expected from *kañiwa.* However, some improvement of temperature is good for all crops and much success was had with raised fields using all the standard crops of the region, such as common potatoes (*Solanum tuberosum)*, *quinua*, beans, barley and oats. The potential of this wonder technology was not in doubt for most early researchers. Today, it may indeed be questioned.

Temperature extremes such as were measured on the raised fields are of particular importance because of a factor not normally commented on and not immediately apparent to the visitor; the whole region of the altiplano is thermally marginal for agriculture. The length of the growing season declines sharply in a gradient away from the lake, where over 180 days may be achieved, to under 150 days, which is a limit for many temperate latitude crops. It is impossible to find a comparable region now cultivated elsewhere, but it may be noted that a fundamental limitation of extending wheat farming on the Columbia plateau, of the northwest of the United States, from the time of introduction of irrigation in the 1930s and 1940s, was the growing season of 150 days. If the season can be extended by even a few days at each end, greater success for crop agriculture is likely. Additionally, the night-day temperature range is a stress factor for crops, and the raised field technology serves to reduce this range, from figures typically averaging about 15-17 degrees Centigrade, to a more acceptable 12-14 degrees (Figure 3.2).

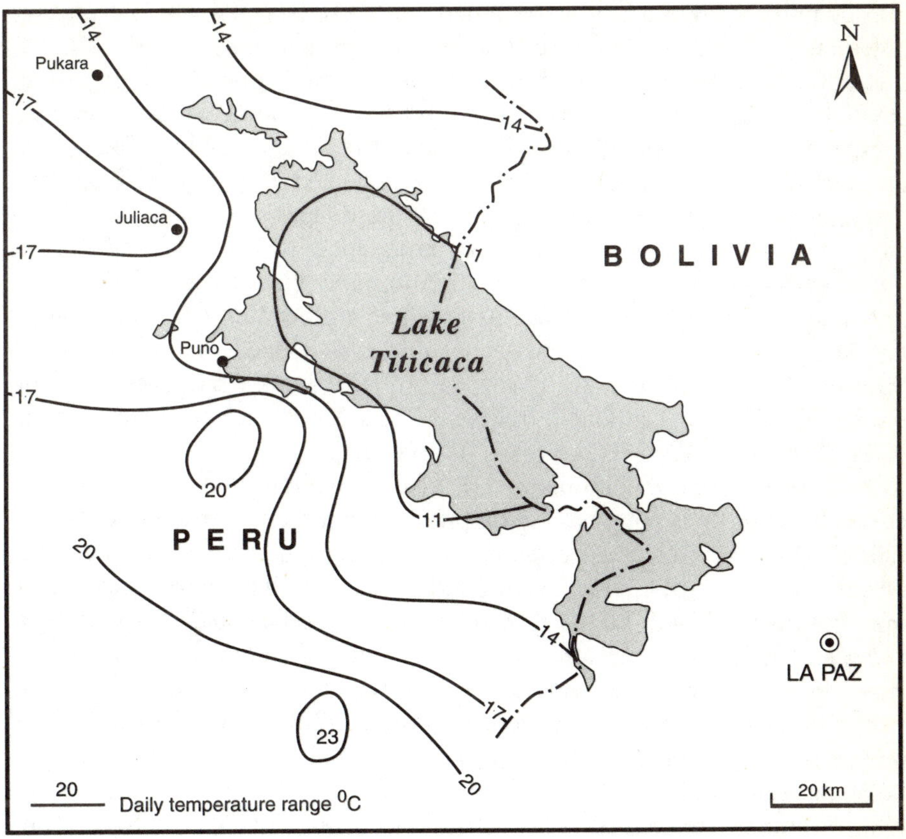

Figure 3.2 **The daily temperature range is a major restriction on crop cultivation in the mountains of the tropics. Even near the lake it is a stress factor for plants.**

Restrictions on Cultivation

Some physical factors work against any kind of agriculture in the altiplano around the lake. Workers for PIWA in trying to establish the best areas for introducing the technology, determined that there were 60,000 hectares of potential land which should be prioritized (Zeballos *et al.* 1994). But of this land, over 53 per cent was liable to excessive flooding, some of it beyond the power of any raised fields to mitigate. Some 57 per cent was subject to clay accumulation in the soil profile. Salt accumulation was a problem in 28 per cent, and poor drainage affected 60 per cent. Much of the land was affected by more than one of these factors. The studies which were used in this report used the experience of previous efforts at introduction, and showed that cultivation in many places would bring salts to the surface, especially in those parts where extra water was brought into the land or where the groundwater table was tapped, as in the areas immediately around the lake. Today, raised field areas which were used in the 1990s and then abandoned, frequently have a thin, and, for plants lethal, layer of salt at the surface as a result of cultivation.

The soils of the altiplano are also sometimes thin. For normal cultivation this does not pose a great problem, but for construction of raised fields, which are formed by taking out the topsoil from the canals and building it up to make an adjacent platform, a very thin topsoil means that the platform itself will be too low to be useful and the canal area will have to be as much as the platform or more. South of Acora, for example, in Peru, there is a layer of calcareous hardpan which is sometimes only 25 cm from the surface. The 25 cm of topsoil turned out of the canals would form only a 50 cm platform, too low for much use, and the canal would have to be 50 per cent of the area used.

Another restriction has been the amount of precipitation. Throughout the altiplano, rainfall is somewhat less than sufficient for crops. South of the equator, the summer period of December-March is the time of rainfall maximum, but the crops planted need rain from time of planting in September, and a dry spring can be calamitous. The effective precipitation compared with the crop requirements can be compared for a typical location in the Peruvian altiplano – see Table 3.1.

Table 3.1 Juliaca – rainfall and crop requirements (measured as potential evapotranspiration through the growing season)

	Sept	Oct	Nov	Dec	Jan	Feb	March
Effective Precip.	17.5	31.9	49.6	69.2	94.4	82.5	75.1
Crop Requiremt.	102.6	104.9	116.1	104.0	80.3	68.5	70.2

Source: p. 346, in Zeballos (1994)

This table shows a rainfall deficit through much of the growing season, but especially in the spring period of September-October. This can be avoided to some extent by leaving the planting until rain arrives, but the growing season is also a major limitation, and except at the lake side, the season is typically only 150 days, marginal for most crops of interest here. A strong recommendation in the study quoted here was thus for much more effort at water management in conjunction with the building of raised fields. If the best results were to be achieved, investment in bringing river water or other supplies was needed, even for the average year. But for most of the region, there is no ready source of additional water supplies. Most of the raised fields have been constructed in areas where the only source of water was rainfall, on the open pampa. In the border lands lying under the hills around the altiplano, there is often spring water or small streams which could be used, but this is the land which is often already in cultivation, and the agencies have been rightly unwilling to disturb an already functioning traditional cultivation in these zones.

A managed supply of water is also needed because of the salinity problem. Highly saline soil, after several years of cultivation, needs to be washed out with a copious supply of water, through a proper drainage system. Most of the projects never contemplated drainage or were unable to do anything about this problem.

Empire or Community: Origins of the Raised Fields

Archaeologists went on from a rapid evaluation of the crop-yielding possibilities of the technology, to argue over the origins of the whole system, an argument which has bearing on the present potential for re-introduction. On the one hand, Alan Kolata supported the idea of empire and organized labour as the agent, and on the other, Clark Erickson saw a local origin and development among humble lakeside communities. The debate, conducted over several years and book length academic treatises, is summarized below. It might be argued that only a society with an organization which we cannot now repeat, such as slavery, could have constructed and maintained the fields, and this would mean that any attempt at reintroduction, especially if founded on the use of traditional methods and non-mechanized agriculture, would be bound to fail. If on the other hand, simple peasant communities were involved, comparable to what we see today in the altiplano, there was great hope for the re-introduction.

Kolata's Model: the Hydraulic Civilization as an Organizing Force for Building the Raised Fields

Kolata's work was conducted in the Pampa Koani, in the valley adjacent to that of Tiwanaku and draining, like it, into Lake Titicaca. He worked with a large, binational multidisciplinary team from 1986, following a pilot project conducted with Oswaldo Rivera of INAR, the Bolivian National Institute for Archaeology, from 1979-82. Work from 1986 focussed on the area around Lukurmata, where a major regional centre with several thousand people had existed in Tiwanaku time. From the small area of raised fields immediately around the site, and the evidence of huge areas further out in the Pampa Koani, where there were only hamlet sized settlements or individual house sites, he deduced that a large subservient population must have been used to provide the city with foodstuffs. This is interpreted as having been part of the Tiwanaku empire system, involving top-down, perhaps coerced labour to provide the food surpluses.

For Kolata, the ancient states used tribute or forced labour from their citizens to fund all their major infrastructural projects, both in the Old World and in the New World. For South America, the logical best example, which he deals with in some detail, is that of the Incas, comparatively well documented by written statements of the colonial period. On the one hand, the Inca divided the lands into three. One part was for the gods, for religious cults (effectively for the state in a theocratic state); a second part was for the Inca himself (effectively also for the state apparatus); and a third part was for the people. Within the third which went to the people, each small community organized the distribution of land in an equitative way to make sure that all were supplied. This system assured the state of control and use of the main land resource. Apart from the land division, there was a labour tribute which assured the Inca of a manpower to cultivate the lands. The labour tribute was paid in work both on the religious lands and on that of the Inca. In addition, a labour tribute was raised as groups of men from every community, the *mit'a*, for major operations with a limited time frame, such as road-building, mining, or service in the Inca army. All the work tribute was raised through the existing locally recognized headmen, the *kurakas*, which made the bureaucracy easier to manage.

Kolata's leap of faith here is to project this system back and to assert that it must have also been the system operated by the Tiwanaku. Because these people were probably Aymara, distinct ethnically and linguistically from the Quechua speaking Inca whose origin was far to the north in the Cuzco region, and since there seems to have been a complete break in between the decline of Tiwanaku and the rise of the Inca, in the north Bolivian heartland, there can be no assumption that they operated in the same manner.

The evolution of the system is argued by Kolata in the following way: in the period from about 800 BC, there is evidence for the first raised fields in Bolivia, whose construction may well have been through the work of small communities or even families, and it is accepted that these, in aggregate, may have been responsible for the construction of large areas of raised field. Each community would have had a local administration, but a hierarchy would have evolved over time, perhaps comparable to what he notes on the north coast of Peru or in Bali; each community is responsible for their own patch, upstream as far as the water source or canal. A higher authority is responsible for organizing the distribution as between rivers or canals, and the highest authorities are at the head of the distribution system.

We might note in passing that even this combination of local control with hierarchy is something which may not have been necessary in the altiplano, since we are not dealing with an irrigation system here, but a system which is dependent on rainwater, or groundwater, with only limited contributions of river water.

After 300 AD, Kolata notes the growth of a centralized state in Tiwanaku, a state which imposes itself on everything, and in the case of agriculture, large scale units are found which he claims were operated by tribute workers. This interpretation is made from examination of the settlement pattern in a relatively small area, a 70 square kilometre area which forms the central part of the Pampa Koani, the valley next north from the Tiwanaku valley. The association of the ancient raised fields he found, with the actual settlements of Tiwanaku time, was difficult because of the almost total lack of pottery remains which are the usual dating tool. But there was a good structural association between the settlements which did all have pottery remains, from the Tiwanaku III and Tiwanaku IV periods (roughly 400-800 AD and 800-1100 AD) and the raised field systems around them (Kolata and Ortloff 1989, p. 237).

At the top of the urban hierarchy, there was first Tiwanaku itself, and then below it, two or three major administrative centres – the main posts of imperial power. One of these was Lukurmata, where Kolata's group did detailed studies. At these centres, there were finely cut stones, brought from a distance, forming the remains of palaces and temples, like a cutdown version of Tiwanaku itself. There were also raised fields adjacent to the intermediate centres such as Lukurmata, but of a special character with much wider benches than elsewhere, and in any case insufficient for the city's population of perhaps 20,000. Below them, nothing but an array of small settlements of hamlet size or less, with an assemblage of remains demonstrating considerable poverty. Kolata also requires the intervention of the organized central state because of the number of larger features, some canals and causeways which involved greater effort than the raised fields. One particularly massive effort must have been in the canalization of the River Catari, which drains the plain around Pampa Koani, and which was straightened in order to reduce its flooding and allow proper operation of the raised fields (Kolata 1982).

In effect, he sees the expansion of the raised fields and their ultimate great success as being linked to their use under a form of despotism which allowed massive production of surpluses to maintain the cities and trade. Ultimately, this argument can be reduced to a kind of environmental determinism, as the control of one natural resource, water, is only possible with one kind of social system, the oriental despotism of Karl Wittfogel (1957).

Before we turn to the counter-arguments of Erickson, we may note that a neighbouring area, the Tiwanaku valley which lies southwest of Pampa Koani, does not show the settlement pattern of Kolata. Instead, the area of the Tiwanaku valley itself, according to extensive studies by Albarracin-Jordan (1999) was occupied by a regular settlement hierarchy, with towns and villages of all sizes. Albarracin-Jordan contrasts his schema of an "integrated nested hierarchy" with Kolata's model of "centralised bureaucracy". While Albarracin recognizes that the major engineering works must have relied on massive labour forces, he reckons that these could have been summoned up in the region for occasional work on the causeways and canals, while for most of the time, each village worked at its own system of raised fields and other structures.

Erickson's Model: Gradual Evolution of an Andean Technology and Work by Small Communities

Erickson's argument on the social organization of production is quite distinct. Working on the northwest side of the lake, he completed a number of precisely recorded archaeological digs through the deposits associated with raised fields in the area of Huatta. The evidence he amasses is of a system which grew up with small family units, each distinctive in style and not showing any evidence of standardization or massive scale. This technology did not require central despots to control it.

The distinctive feature throughout the archaeological digs is variety and change. At Kaminaqa, near the lake, there were a set of round, enclosed blocks of raised fields, with high bunds around them, whether to keep out the relatively saline lake water or to keep out flood water from the rivers. At this site, there were changes in dimensions from one phase of construction to the next, with a change in the wavelength of the fields (the distance between one ridge top and the next). There was change even in the overall structure, with some long, undammed fields replacing the round enclosed blocks in places. At another site, Jucchata, there is evidence for three phases, each with its own wavelength. Phase I had 4.5 metre wavelength, Phase II had nine metres, and in Phase III, the whole level of the fields was raised.

Throughout the evidence is of a gradual building up over many years, of a complex set of structures which varies through time and space. This would not involve massive expenditure of labour at any one time, as the work would be added to from year to year, and from generation to generation. The material of the field surfaces must have been considerable, but Erickson (pp. 140-42) believes that the excavation from the canals would never have provided enough on some of the thin soils. This meant that annually, sediment was dredged from the canals to add to it and to add nutrients; wind deposits would bring more materials, and animal dung would add further nourishment and bulk.

Erickson also uses the evidence for early raised fields, as a marker for the kind of development which took place. He notes there is evidence that the fields were begun as early as 1000 BC, which predates the build up of a substantial dense population. In contrast to a technological advance in the manner theorized by Ester Boserup (1955), where the pressure of population forms a need for technological advance, he suggests a series of gradual advances in a time when there was no pressure. This time could be around 1000 BC. Agriculture around the lake dates back to about 3000 BC, so that there had already been a long time of learning about crops and farming, when the first raised fields were introduced.

The people who made the introduction were probably the Uru, the oldest settled inhabitants of the region, who had learnt to combine the resources of the lake itself with farming (see also Albarracin-Jordan, *op cit.* Ch. 7, pp. 81-90). They gathered the totora reeds which could be used for food, for bedding, for livestock feed, for roofing, and famously for the reed boats of the lake. They caught fish in the highly productive reed areas where the fish spawning areas are found. In low water times, they could use the exposed lake bed, where rich sediments were trapped by the weed, for planting their crops and obtaining high yields. Erickson follows the lead of Karl Sauer, who always maintained the origins of settled agriculture in the tropics to have been at the water's edge, in areas of high potential and a multiplicity of available crop and animal species.

In this favoured environment, the farmers would have observed the best crops could often be obtained from natural raised fields, the levees along the river edges and the edge of abandoned meanders in the floodplain near the lake. Here the first manmade fields could be a simple extension of a natural ridge, and with a growing population be amplified by constructing further ridges.

Later on, from about 200 BC, the evidence of urban concentrations such as Pukara on the Peruvian side, and Tiwanaku on the south side, suggest that the land was worked to produce surpluses and a more hierarchical structure put in place, but Erickson insists that this was never necessary for the early development of the *waru waru* and their extension over large areas.

Albarracin-Jordan (1999) adds to Erickson's ideas about the gradual emergence of the raised fields, the concept that there was a continuity of their use after the Tiwanaku's demise, in the eleventh century. The Pacajes period, from Tiwanaku till the Inca invasion of the fifteenth century, was one in which raised fields are found in direct association with settlements, as well as *qochas* and terraces. His concluding statement on the subject is:

> Knowledge about agricultural technologies was not lost with Tiwanaku's disintegration, demonstrating that this kind of cultural competence was not tied to a centralized planning apparatus, composed of 'hydraulic engineers'. Instead, it rested on the hands of local farmers (Albarracin, op cit. p. 87).

In effect, the argument is that rather than the raised fields being a creation of the despotic state, their technology, and their use, grew over many centuries, and continued on through the empire period of Tiwanaku and on into the post-Tiwanaku period.

In effect, both Erickson's and Kolata's views on the social organization of production can be accommodated. They both agree on the likelihood of early origins of the raised fields, possibly through experiments of the lakeside dwellers. It is likely that on the Bolivian side, or at least in the Pampa Koani area, close control by the Tiwanaku state imposed itself heavily at a later stage. From AD 300 onwards in this area, well organized production with surpluses that were taken by the state, is likely to have been the case in the Pampa Koani rather than small family or community organizations. On the Peruvian side however, at a greater distance and with less close control, the small individual unit may have persisted better, as the archaeological evidence suggests. The important point however is that there seems to have been no necessity for massive state organization, and that a huge area of raised fields could have been built up, as seems likely, by gradual accretion with the effort of many generations of small communities. The physical fact that raised fields do not form a single integrated system, as do irrigation systems with their need for control at higher and higher points on the network, means that central control was never a requirement for success.

A further footnote should be made with respect to early origins. The enormous spread of raised fields in various parts of South and Central America, as compared to only a few isolated finds in the Old World, hints at cultural dispersal of the idea or the technology, rather than separate inventions. This point is left carefully aside by archaeologists, possibly because it has to remain speculative. Many features of indigenous American culture can be traced through all the higher civilizations, such as sunken courtyards, art and architectural styles, and the magical or religious properties ascribed to some birds and snakes. It seems likely that raised fields did not rely on an independent invention here in the Titicaca basin; what is certain is that their function was distinct here because of the unique climatic conditions.

The End of Tiwanaku

On this topic too, there was a difference of opinion between the two groups, those of Kolata and Erickson, and amplified by some other views which seek special causes of a state collapse. The debate is between views which ascribe the downfall of Tiwanaku to social and political breakdown, alternatives which look for an outside cause, some kind of calamity which overtook the empire and ruined it, and a third group which sees continuity over time rather than any sudden end. Again, this debate is instructive not solely as a piece of history, but as a marker for the potential threats which might affect modern introductions.

Collapse Through Environmental Factors or Invasion

Initially Kolata's idea was of some kind of internal state collapse, which led to the abandonment of Tiwanaku and total loss of its knowledge system (Kolata 1982). Earlier writers, notably Posnansky, had also seen what seemed like a sudden collapse, an end to the civilization which could not be explained in terms of the rise of other competitor nations, invasions, or political change, and which must thus be ascribed to a natural disaster.

This left open the question as to how and why the state could have suddenly collapsed. Later, he found good evidence for a possible cause, the powerful changes in the climate of the Altiplano. His theory (Kolata 1992, Ch. 8, pp. 282-302), is that Tiwanaku was overcome by a physical calamity, a long term drought which destroyed the agricultural base of the civilization. He reports the work of glacial geologists working on the Quelccaya glacier. This glacier lies some 100 km to the north of the lake, in the mountains between the Titicaca lake plain and the basin of Cuzco, and its ice record may be taken as an indicator of climatic conditions over a long period. What he notes and describes is a gradual decline of precipitation, as measured by ice cores which can be interpreted as the record of snow accumulation year on year, from the peak around 950 AD. The decline continues through the eleventh century and precipitation is low until after 1400 AD. In terms of actual level of rainfall, it seems (Kolata *op. cit.* p. 290) that there was an overall decline of about 25 per cent, from the wet average of the tenth century to the dry average of the twelfth and thirteenth centuries. This is not a huge amount, but in an area which was already in a rainfall deficit situation, it may have been critical. The ice core data cited by Kolata are supported by sediment cores his team recovered from the lake, showing a change towards deep water flora to those of shallow water over this same period of climatic dessication, as the lake level declined. Kolata's views are followed by other scientists who have examined the ice core data and sediment record (Shimada *et al.*).

The effect on agriculture is judged to have been massive, causing the gradual decline and abandonment of many of the raised fields. The decline would not have been immediate, especially in the areas near Tiwanaku, as they rely in good part on springs which supply groundwater, and ponds (the *qochas)* which are entirely groundwater fed. In some areas, such as the Catari valley north of Tiwanaku, there is a massive canal system which supplied surface water, but even in this case, the canal itself was supplied in part from springs fed by underwater sources. Over a period of fifty or a hundred years, a gradual decline would be felt and the raised fields and ponds become no longer workable.

The evidence of the raised fields is supported by archaeological evidence of the way settlements changed in this area after 1000 AD, at least in the Pampa Koani area where Kolata worked. While before 1000 AD, there was a well organized hierarchy of settlements with villages and major cities, after that time the settlements of the rural area are tiny and dispersed, and the cities are apparently unoccupied. The evidence also suggests a move towards more stockrearing of llamas and alpacas, and away from cultivation altogether.

Another version of the collapse of Tiwanaku is that calling on evidence for an invasion of the region, by outsiders who destroyed the empire. Albarracin-Jordan summarises these arguments; central is the idea that the original inhabitants were the Uru, the people who are now confined to fishing around the lake, and that these people, or the Pukina who followed them, were the Tiwanaku people, and were displaced by an invasion of the Aymara, in the period before the Incas. This invasion would have been responsible for the destruction of the raised fields and other developed infrastructures for agriculture. There is some supporting matter for this version of events, from the Spanish colonial writer, Cieza de Leon, who collected information in Peru on the pre-colonial history of the peoples around the lake. But the case made is still flimsy, and unsupported by other kinds of evidence.

Graffam (1992) also reports a variant on the collapse being due to destruction by enemies. In this case, those responsible were the people themselves, possibly disaffected and oppressed inhabitants of the peripheral areas of an over-expanding empire. This is a view of state collapse held by some Latin American archaeologists, evidenced in the destruction of some of the peripheral settlements of the Tiwanaku empire. The result was disintegration of the empire into a series of local chiefdoms or regional kingdoms, called the Aymara Señorios (Lordships) which came into being as Tiwanaku failed. The Lupaka and the Qolla were the two most important of these on the northwest side of the lake. On the other hand, Graffam also makes the critical remark that these uprisings, if they took place, were not necessarily the cause, and may have been the effect, of a state collapse.

Gradualism in Change

On the northwest side of the lake, Erickson does not find evidence for a collapse and abandonment of the raised fields. Instead, around 1000 AD (Erickson 1996, pp. 154-56), there is a second phase of expansion and rebuilding of the fields, known as Phase II, which continues possibly to the time of invasion of the Incas, around 1450 AD. This is completely at odds with the findings of Kolata, and suggests that different conditions were obtained. In fact, Erickson believes that the idea of collapse is inaccurate for the events on the northwest side, and that instead, after the period up to 1000 AD, there was a change to local ownership of the fields, and then a gradual decline when first the Inca people invaded, then the Spanish. The later, Phase II raised fields seem to be associated, no longer with the Tiwanaku, but with the Aymara lordships mentioned above. The evidence on total area of raised fields and density of settlements which Erickson found around Huatta, shows that farming was intensive and successful at this time. Politically, what seems to have happened is a transfer of power from the Tiwanaku empire, to a set of regional kingdoms, which maintained this power until the Inka took over.

In a later paper (Erickson 1999), Erickson expands on his arguments against sudden change, even if there were major climatic variations to cope with. Climate in the Titicaca basin is always subject to major changes (Erickson points to the pioneering work of Monheim on this topic) and peasant populations have adopted coping strategies for dealing with major droughts and flood periods. One way of coping is the building of what has been termed 'landscape capital', in the walls, fields, terraces, and drainage systems of the land, and this has been done successfully over a huge period on the lake plains of the region. Erickson remarks that floods usually follow droughts, and that some structures of the landscape are in place to cater for both of these. The raised fields were not in use all the time, and when there was low water, new lake mud of high fertility was uncovered with the possibility of raised fields construction to reduce waterlogging. When there was flooding and high water, these lands would be under water and other, higher lands would be in use for their existing raised fields which would rise above the floods. A fall of only one metre in lake level can expose an estimated 200,000 ha. of previously submerged lake bed, and man's occupation and use of the lakeside plain must therefore be flexible enough to cope with huge changes in the land resource. Cold water is effectively poured on the idea of an environmental determinism which would see the fate of the empire and of its raised fields controlled by the fluctuations of the lake and its climate.

On the Bolivian side, Erickson's view is supported by the work of Albarracin-Jordan (1999). This writer reports the extensive archaeological work done in the Tiwanaku valley, where a total of 964 sites are recorded from both Tiwanaku and post-Tiwanaku time, and which mostly show a continuity of occupation from one period to the next. One difference is that the post-Tiwanaku settlement, labelled the Pacajes culture, reaches up to higher altitudinal levels, and another is the presence of some forts, indicating an element of conflict between different groups after the decline of central power at Tiwanaku itself. Albarracin also comments on the relationship between the Pacajes sites and remnants of raised fields, terraces, and *qochas*, indicating that there had been no loss of the agricultural knowledge among these people despite loss of central power. Various other indicators are noted by Albarracin, including the ceramics left by the Pacajes people, the funerary customs, which all show continuity rather than abrupt and drastic change.

One of the workers with Kolata also challenges the idea of any dramatic change in the raised fields, whatever the changes in political organization (Graffam 1992). There are a number of strands to Graffam's argument, but the essence of it is that a continued high yielding agriculture, such as that from the raised fields, was needed by altiplano people who maintained huge herds of camelids, both alpaca and llama. Graffam's calculation is that 1.8 to 2.8 million of the camelids were maintained at the time of the Incas, and maintained by the same communities that conducted intensive agriculture with raised fields. These herds were a drain on subsistence farming, so the raised fields were needed to subsidize pastoralism. The post-Tiwanaku people were also able to maintain and rebuild the raised fields, as it was an efficient use of their time and energy, according to Graffam's calculations of cost.

These arguments seem too theoretical to be a strong support to his claims, but of greater strength is the evidence from archaeological research which he conducted, directly in the Pampa Koani, where Kolata's thesis is based. Here Graffam shows that there was a network of small settlements or individual house mounds, in post-Tiwanaku time, and that these are associated both in topography and ceramic remains, with raised fields, many of the latter built over previous ones from Tiwanaku time or previously. His conclusion is that this was a community based technology, robust enough to withstand any political change and continued on up to the time of the Incas.

Inca Invasion

Abandonment at the time of the Inca invasion is possible to identify, and is relatively easy to interpret. Erickson (pp. 269-76) recalls the Mitma system of the Incas, exchanging peoples from one area with another to avoid ethnic concentrations which could be dangerous. He also notes the effect of wars and disturbances, which would easily break down the specialized agriculture of the raised fields. Incas are also likely to have deliberately destroyed local farming systems along with local social structure, to avoid building any local power base which could threaten them. There were also cultural differences, notably the Inca emphasis on the use of terraces, and growing of maize for consumption and especially as a crop with religious significance, which did not fit with the use of the raised fields. Terraces had been used by the Aymara peoples, but the Incas had a still greater focus on them. One additional reason he does not mention is that the raised fields would not have been understood well by the Incas, coming from a better watered zone, with milder temperatures. The very uniqueness of the system would have made it difficult for an invader to assimilate. This point could be made with still greater force in reference to the Spanish, who understood ordinary irrigation systems, but for whom any raised fields would be irrelevant to agriculture.

Overall, the argument from Erickson and from Albarracin is that, in contrast to Kolata's climatic disaster, there are social or political controls over the structure of agriculture, first bringing the changes from Tiwanaku to the Senorios, and then to the Incas and finally the Spanish. The message of the gradualist view of change in the altiplano, is that the raised fields, and other elements of Andean technology, were not dependent on the existence of a dictatorial empire or some similar form of absolutism. Instead, these structures were capable of being maintained over substantial changes, even in troubled times, as features of indigenous knowledge that could be readily transmitted.

This argument does not question the reality of climatic change, which is likely to have had major effects on the organization of the state in prehistoric times. Some writers have chosen to question the climatic change, for example Denevan (2001), because the collapse of Tiwanaku is not well documented and because the dates for an onset of drought are not clear. However, it is clear that major climatic changes are possible in the region, induced by the same forces that cause the *El Niño* effect on the coast, and that in some respects it is a marginal region, susceptible of devastating blows if drought or flood take over.

It is worth noting that there is outside corroboration of the idea of a climatic change of some magnitude forcing the onset and the end of major civilizations in the continent. Reid Bryson, climatologist of the University of Wisconsin, made a comparison of the beginning and end dates of major civilizations in South America, with major climatic change dates (Bryson 1978). His findings are of interest: while the random expectation would be that about a third of the civilization limits would fall within 100 years of climatic change dates, in fact about 74 per cent of these dates do coincide. Including more precise dates for Tiwanaku which he did not have, this level rises to over 81 per cent. In other words, climatic shifts seem to have a relation with the rise and fall of major cultures that cannot be chance.

Conclusions

Research in the past forty years has in sum been highly positive, in relation to a possible reintroduction of the raised fields technology. In the first place, the physical effects of the use of raised fields have been well demonstrated, especially the microclimate effects which are so important to the survival of a crop in this physically hostile region. I have raised a question over the lack of long term work on the effects on soils, which may have been an Achilles heel in the research. Secondly, it is shown that at least it was possible, and probably quite true in fact, that simple communities of farmers could have developed the technology over long periods of time. Regarding the sudden disappearance of the technology and its total loss from folk memory, it is seen that major climatic events may have had disastrous effects, especially a very long term drought, and that invaders with no concern for maintaining the technology may also be responsible for its disappearance. In other words, this was a sustainable way of farming, one which could be maintained successfully over long periods of time. Such was the setting for the wave of projects to bring this long-forgotten system back to life.

4 The Projects for Reconstruction

Introduction

From the early 1980s, a host of programmes were put into operation, born of the enthusiasm of discovery and sometimes ideological conviction, to bring back the raised field technology to the lands around the lake. Many of these achieved very little and were soon forgotten; a very few have remained operative to the present day, though usually with substantial modifications from experience. The context, to anyone coming from the outside, was never promising. There had been a failure already of too many comparable projects in the rural areas of Peru and Bolivia, focussing on a single innovation, after only a few years of enthusiasm. These included the programmes for *huertos familiares* or family vegetable gardens, for biofertilizers, biogas, water pumps, windmills, solar energy, well perforations, guinea pigs, and rabbits. One of the few which maintain a visible presence in the landscape, was the programme for greenhouses, which were advocated by a number of agencies and regional government, both in Bolivia and Peru. They were made of heavy plastic sheets, hung on a light wooden frame over a base built up of low adobe walls. Today, the adobe walls remain, but the plastic roofs, if anything is left of them, are usually in tatters, dried and perished under the intense ultraviolet radiation of the high altitude plain. There are no crops left in these remnant greenhouses, though sometimes pigs or chickens are kept in them. Those that have had their plastic roofs renewed are mostly on state experimental farms and similar institutions. Any kind of arable farming project might have been regarded with suspicion, for most of the land, and most of the farm income, is in livestock farming, not cultivation. In Puno Departamento, the main area of raised fields in Peru, only 3.6 per cent of the land is reckoned to be arable, and only 2 per cent actually cultivated in any one year (PIWA 1992a).

Against all these negatives, there were many attractions in the idea of the raised fields. Firstly, they were in line with current thinking of the 1980s which was highly critical of the Green Revolution of the 1960s, a

2. A raised field with bitter potatoes, in a project community run by PIWA, at Pomata. The PIWA agent, Ernesto Chura, accompanied and helped the author in the field.

3. The author at one of the earliest raised fields projects near Huatta, now abandoned.

4. Harvest time at Huatta, near Puno. Barley has been cut by hand and laid out to dry before putting into stooks. All harvest work is by hand and in this system, the reintroduction of an ancient technology might seem a godsend, allowing more production without recourse to modern machinery.

revolution based on high technology genetics and use of heavy inputs of water, fertilizers, and chemicals. Raised fields would be a viable low technology alternative. Secondly, the fields gave the promise of massive and increasing production and productivity of the bleak high plateaus, unlocking the secret of how to make these lands bloom. A third important set of reasons for enthusiasm was in the apparent independence of the technology from foreign inputs, such as capital, machinery, special seeds. This could be a sustainable development programme, at a time when sustainability was entering the development vocabulary. If projects did not need outside inputs then they must be sustainable. If they had worked for centuries before, they had already proved their sustainability in this sense. And although just how they worked had been long forgotten, raised fields could be regarded as coming from indigenous knowledge, which was to be a privileged concept in the post-modern development era. A fourth kind of reason, particularly powerful in Bolivia, was that the whole idea could be construed as nativist or nationalist. In Bolivia, the Tiwanaku people were thought to be the originators of the technology, and they were possibly the same as the Aymara speaking people. Today, the Aymara nationalist movement within Bolivia, for cultural autonomy and the recovery of ancient Aymara language and political status, is gaining rapidly.

Given the poor record of the previous single-item agricultural programmes, one could be forgiven for being cautious about the likely results. But the attractions were what impressed many, and major programmes were instated while the enthusiasm was still powerful. There was great diversity in outcomes and in the kinds of problems encountered. Programmes were led by different kinds of organizations, from both the public and private sectors, and with different aims in mind. They also had to deal with varied kinds of rural communities whose attitudes to any innovation would be unpredictable and often quite negative. There was a learning process for both farmers and for institutions, and even the aims of the projects changed over time. Differences emerged too, between the developments in Bolivia and Peru, based on the physical conditions, the social and economic structures and the varying forms of land tenure.

History of the Peruvian Programmes

The first attempt to transfer the raised fields from the distant past to the present day farmers, was that made in connection with the academic studies by the archaeologist, Clark Erickson, in Peru. My first introduction to the subject was by taking an ancient bus on a dirt road, filled with farmers and their families, going out from Juliaca with their purchases, vegetables, cloth for making clothes, pigs and chickens in sacks, together with a little cash from their own sales in the town. We were going out to the town where Erickson had worked, Huatta, a village masquerading as town, sitting picturesquely on the edge of an island of hard rock, some two miles long, in the middle of the Titicaca lake plains. All round this long ridge of high ground, a dozen hamlets lie on the break of slope between hill land and flat plain. This is the typical site for villages of the altiplano, and suits farmer needs, because it gives good access both to the cultivated patches on the slopes, and to the plain below which is still largely used for grazing animals. The first experiments were taken up with the farmers who inhabit these hamlets. They made particular sense, because throughout this zone there are thousands of raised field ridges from the past, showing that this must have been one of the major concentrations of this technology. As noted in the last chapter, Erickson's work extended to more than the experimental archaeological work of reconstruction, and included theorizing on the origins of the raised fields, their relation to local cultures, and archaeological digs in the area to trace and date the records of construction (Erickson 1996). To assess the ways in which the raised fields, visible only as traces in the plain, operated in prehistory, he used experimental archaeology to try to recreate some of the structures and functions, on the same locations and following the same patterns as the original fields. He involved local labour in ways as near as could be conjectured to those of the original constructors, using non-mechanized labour using traditional tools. His work extended over the period 1981 to 1986, some of it in conjunction with a postgraduate student, Ignacio Garaycochea, who came to live in the rural area and personally direct the work of reconstruction. Erickson's reconstruction work affected some 2.1 hectares, while the extensions by Garaycochea added some 14 hectares more, all in the area around Huatta. The work expanded out from this purely academic, archaeological experiment, to include the establishment of a project, *Proyecto Arqueológico de los campos elevados*, or PACE, whose function was principally to diffuse the information about the raised fields to other institutions and interest them in further projects.

This led directly to project studies set up by the local university at Puno, the Universidad Nacional del Altiplano (UNA), which made a photographic survey of the traces of raised fields in the area, and to interest by the nearest agricultural research station, at a site between Puno and Juliaca called Illpa, the Instituto Nacional de Investigación Agropecuaria y Agroindustrial (INIAA), financed by the Canadian Agency for International Development, which studied the kinds of crops suited to the raised fields and the diseases and other factors which might affect crop yields. Other governmental backed bodies joined in this enterprise such as CORPUNO a governmental initiative for bringing capital into farming (in existence since the 1970s) and the Micro Region Puno, another Peruvian central government initiative of the same kind with regional development aims. Others were Ministry of Agriculture programmes, CIPA, PISA, and CESPAC. Religious entities were attracted into the business, both the Catholic Church (C.A.J. Maruri) and the Adventist mission (OFASA). What was absent generally was the normal kind of Non-Governmental Organisation (NGO) with foreign financing and represented by small teams of foreigners anxious to set up projects to alleviate poverty. Those that did attempt projects with the raised fields had only a brief presence for one annual campaign or two, and they did not survive. This could be regarded as an advantage, because the ongoing programmes for reconstruction were undertaken by skilled personnel with experience of working in the altiplano conditions and of attempting rural development among farming communities. A central study was that of the potential area for reintroduction of raised fields – some 82,000 hectares had been reckoned by Denevan, but 102,440 hectares were calculated from aerial photography by the Puno University study (Diaz and Velasquez 1990) which if ever implemented in full and as expected over time, would have meant a doubling of the total cultivated area, plus its conversion from occasional low yielding crops, to highly intensive raised fields with massive production. In fact, at least three calculations of the potential areas for reintroduction were made in Peru, and two in Bolivia; it was obviously thought necessary to plan for a massive programme of reintroduction, and many thousands of hectares were envisaged by the enthusiasts in the main agencies.

The projects also benefited from a notable concentration of efforts, most of the work being done in the area around Huatta, where Erickson had begun. All the early projects were within a few miles of this site (see Figure 4.1). Notably, the efforts organized by the experimental station INIAA, with a local development agency, CIID, and funded by the Canadian Agency for International Development, were focussed in the same area as begun by Erickson. This must also be regarded as an advantage, for the information and positive attitudes towards the projects would spread from one community to another with some ease. Later ones would be much more spread out (Figure 4.2).

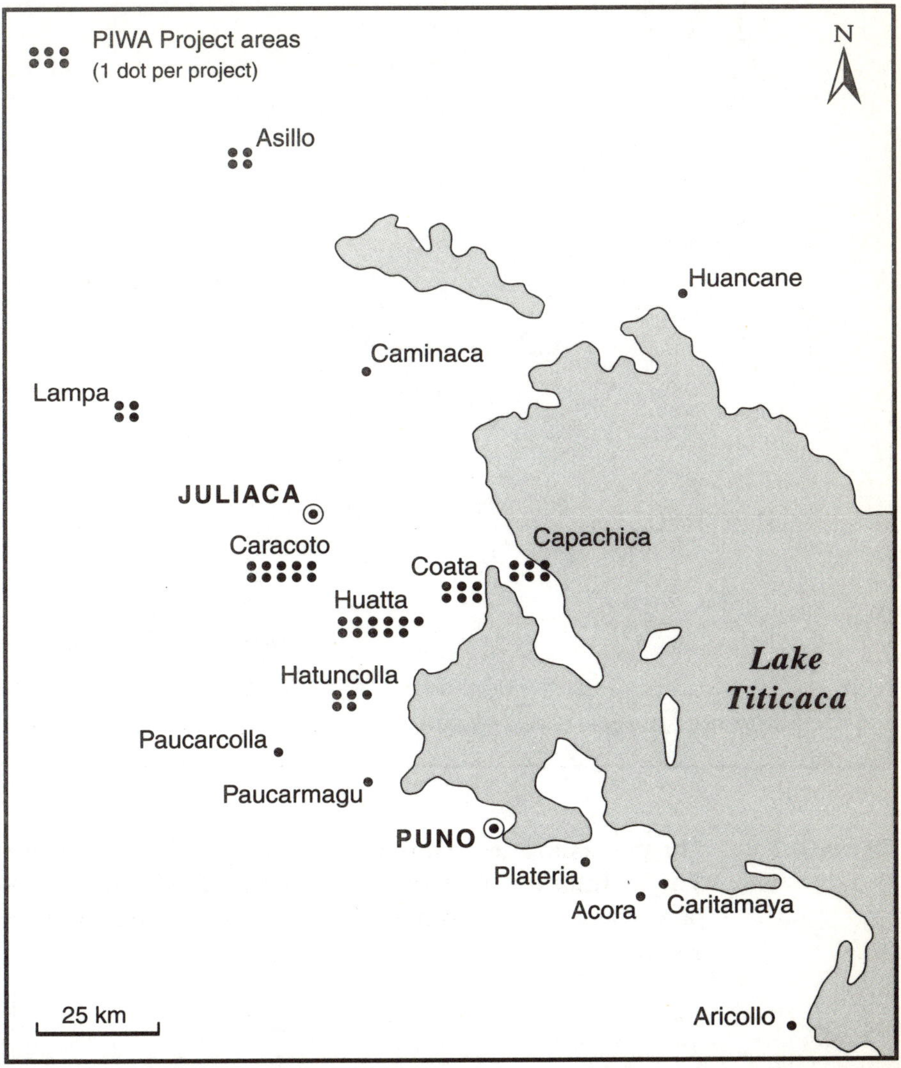

Figure 4.1 This shows the early project areas up to about 1990. They cluster heavily in the core area around Huatta.

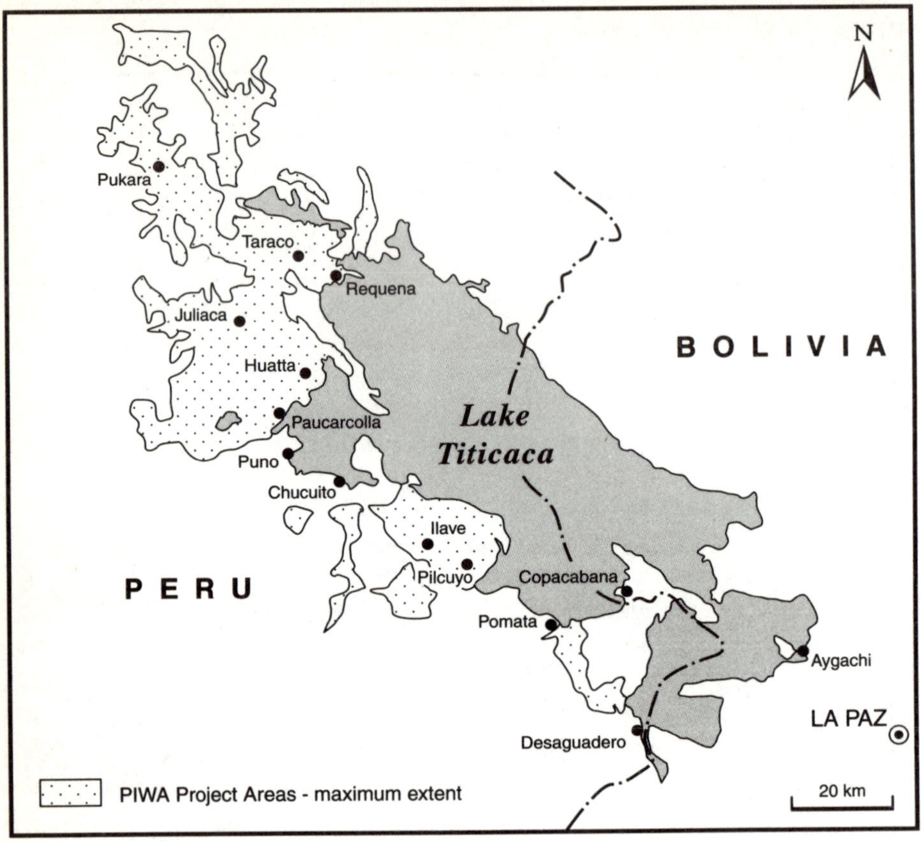

Figure 4.2 **The prehistoric raised fields area around Lake Titicaca. The estimates of Denevan, based closely on aerial photography, are much more modest than those of Kolata.**

From 1989, an organization with funding from the Binational Agency for the Lake Titicaca, PELT, was set up with offices in Puno, and this agency PIWA, The Inter-institutional Progamme for Waru Waru, has been prominent in published and unpublished research, as well as in active promotion of the raised fields. It is worth pointing out that the agency was given the direct remit to conduct research on the raised fields and to promote the development of this technology (Villena 1996). It was, like the other agencies, always committed to a positive drive to incorporate land into the new technology. Its work is officially documented in its annual reports to main offices in Lima of the Swiss Development Corporation, COSUDE (PELT-COSUDE 2001). PIWA has claimed to reach a universe of 6400 families, and have completed 4000 hectares of raised fields. One of the main tasks undertaken by PIWA was the estimation of the total area which might be put into raised fields. A series of studies were done in the early 1990s, the first giving 102,000 hectares of potential raised fields, the second 142,000 hectares. This huge area was eventually decided to be too great, and it was noted that at the rate of building being achieved – 1000 hectares over ten years – it would take a thousand years to reach something approaching the target. The third estimate based on a selection of areas on both physical and social characteristics, was reduced to 62,000 hectares. Even this target now looks to have been hopelessly optimistic: at the end of PIWA's programmes, the area now operated as raised field is at most 300 hectares, and in decline. From reports seen in Lima, at the COSUDE headquarters there, the amounts done every year up to the end did not exceed 100 hectares by much in any year up to the end of the programme.

PIWA operated over a four phase programme, starting in 1989-92 with Phase I, continuing in 1992-95 with Phase II, and with Phase III from 1995 to 1998, Phase IV lasting to 2002. The first two phases were ones of rapid expansion, without close control or field testing, and with PIWA acting as general centre of administration and guidance for a set of about fifteen other institutions which were engaged on raised field projects. The later phases were more organized and used more research as a basis. They also moved away from the early emphasis on community projects, which were found to be conflictive, towards projects based on a family or group of families. A major change was away from any money incentives, to provide only advice and technical knowledge, with the possible addition of seed for alfalfa. PIWA remains to the present, renamed PIWANDES to reflect what is in effect a new organisation with a new remit. With the end of the raised field programme, the team has been disbanded and the director has formed a new core group in order to formulate different kinds of developmental projects based on a wider range of rural aid, including areas away from the altiplano.

Apart from PIWA, the only agency which remained into the 21st century was CARE International, the international development agency which started work after World War II and was based in Canada. Its funding is that of CARE itself, plus the Dutch government. CARE was given a broader remit than PIWA, and was always able to link its projects to such things as marketing and training in general, whereas PIWA only attempted this at the end. CARE also focussed on the lakeside communities more, while PIWA attempted many projects well further inland and onto the *suni* land which experiences a harsher climate. CARE, like PIWA, changed its focus in 2002 although its region of concern remains the altiplano, and it now seeks to promote small river basin development based on ecological principles.

The *modus operandi* of these main agencies was modified from that of the original projects, and became more sophisticated, though there are many elements in common. Neither CARE nor PIWA offered money to the communities involved in the later years, nor was there food aid, tools or other material incentives, apart from the use of alfalfa seed, to those communities already involved in projects, for seeding and restoring the nutrient levels in old and abandoned *waru waru.* Alfalfa seed was recognized as being a key element, late in the day, because of the huge problem of trying to conserve some fertility in the soil.

In the early projects, the locations for rebuilding the raised fields were picked out by the agency, and then the local population told of the project, very much in top-down form. This has changed over time, as PIWA became aware of the need for participation by local people. CARE was careful not to initiate projects, relying on local concerns and on representatives coming first to the CARE office in Puno in order to seek help. For both agencies, the recent projects depended on good local interest, since no money advances were available after about 1995. Too many other programmes had relied on using money payments for work and seed, and had engendered a kind of dependency on agency charity, a situation called *asistencialismo* in Latin America.

In the case of CARE, as mentioned, there was a broadening out of the project nature and a seeking to link the production system of the raised fields, with some processing of the products and efforts towards marketing of goods. According to Alipio Canahua, the director of research at CARE's Puno office, the aim has been to give small groups of people the power to control their own economy, and this can only be achieved by moving down the supply chain, from farmer producers, who have always been weak, to producers and sellers in the final market. The achievements in these linking movements have been limited, but the ideas have taken root and are well received. For PIWA it was not possible to broaden out the programmes, as this agency's remit had been given very narrowly, to introduce the raised fields and promote their extension in the altiplano.

The Bolivian Experiments

In Bolivia the efforts and the organizations involved have been smaller in extent, though no less intense and enthusiastic. A smaller area for projects is understandable, as the area likely to be useful to raised field technology, the flat lakeside plains area, is itself smaller, perhaps one fifth of the Peruvian. The process in Bolivia started with an investigation, similar to that in Peru, into the archaeological meaning of the raised fields, in the context of the Tiwanaku culture. Alan Kolata, of the University of Chicago, commenced research into the Tiwanaku culture in 1981, and became interested in the forms of agriculture suggested by the remains of raised fields. He was funded by the US National Science Foundation, and the National Endowment for the Humanities and a variety of non-governmental funds. His start was slower and more hesitant than that of Erickson, perhaps because the latter had to deal with Quechua-speaking peoples, north of Puno. These people are regarded as relatively outward looking and adaptable compared to the Aymara south of Puno. Aymara are seen as inward-looking and reluctant to adopt anything which compromises their ancient culture system. These are local appreciations. My own view is not of a more withdrawn people, but one which has maintained more clearly its ethnic roots. In the North Bolivian altiplano, most rural families speak Aymara and retain well the use of ancient customs and ceremonies, clothes and other markers of their ethnicity. This has obviously made it easy for any outsider, including the Bolivians of the city, to mark them out as backward. Erickson had, in any case, the advantage of a resident helper, Garaycochea, who lived in Huatta for years until the fields had been established.

Kolata (1996) records the difficulties in negotiating over several years of community meetings with the Aymara in upland Bolivia, before agreement was made with the people of the small town of Lakaya, northeast of Tiwanaku. He had hoped to establish the raised fields as a grassroots venture, with local initiative; but in the end the project agreement was based on a promise of food aid, in return for allowing the use of the flat pampas for the fields. This stepping down from an early ideal is significant, and notable in the context of later experiments and projects.

Eventually, he reconstructed some four hectares of ancient raised fields at Lakaya in 1987 (Kolata 1993, Kolata and Ortloff 1989). The results were just as spectacular as those first found in Peru. In a sharp frost on the nights of February 28-29 1988, there was a killing frost with temperatures as low as -5C on the open pampa at Lakaya. Crops on the open pampa of potatoes and quinoa were severely damaged, and crop losses of 80-90 per cent occurred. The raised fields at Lakaya suffered only a limited frost burn of the leaves and a crop loss of about ten per cent (Kolata 1993, pp 190-92). The same story was told for other crops, of barley, broad beans, quinoa, kañiwa, onions and lettuce. Only a few plants of maize, planted as pure experiment, were lost on these fields.

The success here led to further investment from the University of Chicago, supervised by Kolata, and a major programme termed Wilajawira, named after the Pampa de Wilajahuira (the Red River Plain) where it was set up, led by the local archaeologist, Oswaldo Rivera, later to become Director of the Bolivian National Institute for Archaeology. Wilajawira, later called the Wiñaymarka Foundation, became a standard NGO, funded from the United States and from the Swiss and Dutch governments, and it set up projects for reconstruction in the region, by reconstructing fields in the valley to the south of Pampa Koani, the Tiwanaku valley, where they were built very close to the archaeological site of Tiwanaku. It also built a total of some 130 hectares of fields, including areas at Batallas, and at Copacabana. All these projects have now terminated, and the land in raised fields has been largely abandoned, though other sites nearby have been taken over by other projects. The central site lying as it does adjacent to the ancient monuments of Tiwanaku itself, presents a rather forlorn aspect to the visitor. It is a wetland site, now totally covered over with thick grass, and grazed by a few cattle. It is as though the raised field project has become part of the ancient history represented by the massive stone remains of palaces and pyramids, a mute testimony to past efforts and past knowledge which has already been totally forgotten.

The Wiñaymarka Foundation was the single institution developing raised fields for five or six years, throughout the late 1980s, in contrast to Peru where many groups were involved early on. Its full name became in fact, "The Wiñaymarka Foundation for the recovery of Andean Culture", and this shows a very definite orientation in favour of bringing back various elements of the Aymara culture. It encouraged various annual ceremonies and rituals and positively tried to link these in to the *Suka Kollus*, the raised fields which had become a symbol of the revival of ancient culture.

But eventually its work was taken over by other organizations, and reoriented away from the cultural and towards economic and rural development aims. Much work has been led forward by the Catholic University's Institute for Rural Development in La Paz. From its director, Licenciado Hugo Ossio, I gleaned a picture of the work which has been done, complemented by a number of field trips covering virtually all of the established areas. From about 1989 to 1994, the projects were led by NGOs, such as Plan Internacional and CEDEFOA. The universities in La Paz set up their own projects, including the Superior Military College, (EMI) and the Saint Andrews University (UMSA, or the Universidad Metropolitana de San Andres). Competition and conflict between these meant some initial difficulties, and a slow start. Many projects were dropped after only a year or two, never to be restarted. Evidently, the university has had a major interest in maintaining the projects, as it could use them as part of its own research and its students could write theses based on the field experiments conducted for the programme.

Then in 1994 the Swiss Development Corporation (COSUDE) stepped in to organize and finance the Bolivian programme, and agreements were made between several institutions in Bolivia, resulting in the agency PROSUKO, which has assumed a supervisory role, overseeing the work of both private and public organizations in Bolivia, helped by funding from the Swiss COSUDE. The constituent organizations of PROSUKO are the Catholic University, the St. Andrews University (UMSA), the Superior Military College (EMI), the National Meteorological Service, and PROIMPA, an NGO. An army unit, the eighth Infantry Battalion, has also been associated with the programme, using the projects on its lands as an exercise for conscripts and collaborating with scientific research (Hamilton 2000). The overall direction of PROSUKO has remained in the hands of COSUDE, and its work over the years can be followed in the annual reports held at La Paz by that organization, which are available to genuine researchers.

PROSUKO

PROSUKO's work since 1994 has been more administrative and academic than engaging directly in field projects, most of its research being done through the agricultural experiment stations at Batallas and Tiwanaku, where there are also colleges of the Catholic University, offering intermediate courses for students who will go on to study at the main campus of the university in La Paz. Most of the current projects, from field observation, are clustered in the vicinity of the two experiment stations, The total number of projects is put by PROSUKO at 70 plus, but again, most of these are inactive at the present time. PROSUKO does have a substantial budget, in the 1999-2000 season amounting to 318,000 dollars (COSUDE 2000). Much of this however, goes on the research programme at the agricultural stations, rather than on the projects.

From the various annual reports, the programme adopted by PROSUKO has been somewhat similar to that by PIWA in Peru, following in its steps a few years later. In 1994-95, there was a short period for establishing a structure for the organization and acting as coordinator for many small schemes under NGOs and university agriculture departments. This was followed by a phase of research in 1995-99, involving 16 hectares. In the following third period, from 1999-2003, the effort is transferred to field experiments and extension work to encourage diffusion, and official documents suggest that perhaps some 100 hectares of land are currently affected. In reality, this is likely to be much less. From my own surveys of the areas of main activity, near the Agricultural Extension Stations at Batallas and Tiwanaku, there are less than 90, and possibly only 60 hectares, in use now at any one time. If there are other areas they are remote and unconnected to the principal areas of activity.

PROSUKO was not the originator of raised fields in Bolivia, and so it inherited a collection of different projects and sought to coordinate them, rather than becoming the active agent. In the first years, it had projects in four areas (see Figure 4.3). To the northeast there were small projects at Ancoraimes, a second concentration was around the town of Batallas, a third near Tiwanaku and the original projects of Kolata, and fourth area at

Figure 4.3 **The early plans of PROSUKO were large scale and involved the whole Bolivian shore of the lake.**

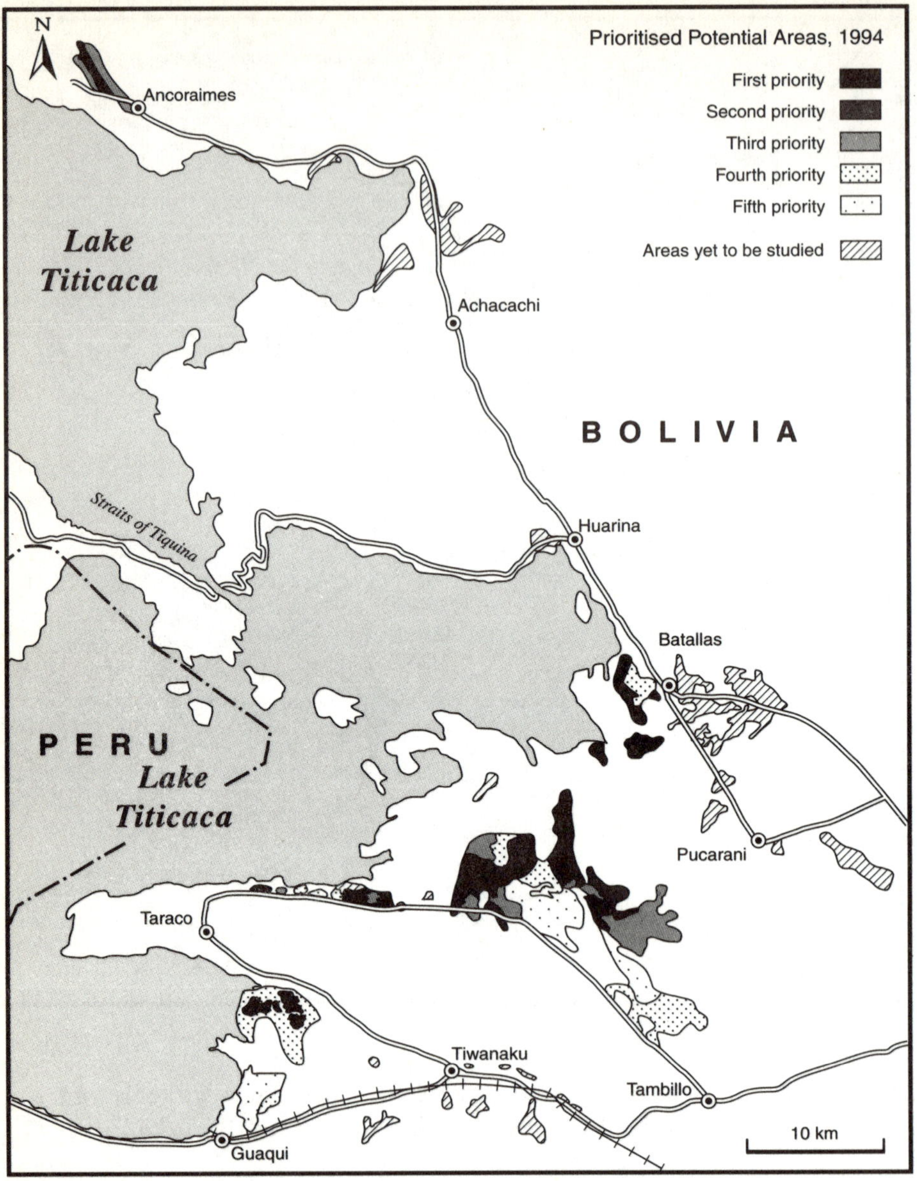

Figure 4.4 **By 1994 the areas had been refined and reduced, but some priority areas never had projects on them.**

Aygachi, further west. After three years, these had been reduced to just two areas, Batallas and Tiwanaku (Figure 4.4).

The others were more remote and difficult to maintain from bases at La Paz, though no full justification was ever given for their abandonment. An internal document (COSUDE 2000A) notes that it was more efficient to concentrate on the two areas because of their higher potential and the positive reactions of local people in these areas. It is certainly the case that some diffusion to non-project farmers has been achieved in these areas, and that more opposition was experienced in the others. The concentration on just two areas would appear to be an acknowledgement of the difficulty of promoting the programme, and the need to work from strengths. Early annual reports spoke of the aim to spread the technology all along the lakeside, but this has been quietly dropped from later reports. The detailed grading of potential areas has been of little value as all depended on the will of local communities and the will to commit time and effort to projects.

PROSUKO has also, belatedly, recognized that the initial focus was too narrow. Its recent annual reports (COSUDE 2000) say that the objective is to increase the *suka kollus*, but also to help production in the open pampas. It also wishes to promote post-harvest technologies (storage of products and processing of foods), and marketing. In this aspect it is following down the line set by CARE. For the first time in this same document (COSUDE 2000, p. 10) there is the mention of the *suka kollus* as a complementary alternative (i.e. not a total answer to the farmer's prayer), and as something which reduces risks and provides diversification.

Organization

On the Peruvian side, the organization of the projects was in the early years commonly through an institution working with groups, typically the recognized *comunidades campesinas* which had been defined and given stronger rights by the Peruvian agrarian reform of the 1960s. Consonant with this was the use of communal land for the projects, and communal work for the maintenance. These were never strict rules, and some projects were carried out on private land, whose owner had agreed to have the project on his land for a number of years. When the agency paid wages for construction, the payment was for so many metres of field constructed; when the community undertook the work, all were supposed to contribute.

After a number of years, conflicts in the communities, failure of all members of communities to turn up for work, or to work a whole day, caused a move towards family or family groups as the target for projects. Another problem was in disputes over the ownership of land and who should claim the benefits of built raised fields. If the project had been built on private land, one private owner was left at the end of the project, commonly a four year term, with fully constructed raised fields, while all other farmers were left with nothing. In the later phases in Peru, while community emphasis was always officially maintained, it became the standard practice to work only with interested smaller groups.

From what I could make out, the decision about the zone in which efforts would be concentrated and the community chosen, depended in large measure on local interest. There were large studies done to determine the most appropriate areas to have projects, and these worked out areas of interest in terms of physical potential, lack of competition with existing cultivation but existence of a cultivating tradition, and availability of abundant labour needing employment (Zeballos 1994). But in the end, the areas actually worked on were so small compared with the potential, that zoning of high and low potential lands never became important.

On the Bolivian side, there were attempts in the early years to follow the Peruvian community pattern and work with the peasant communities, but these proved most intractable. Under PROSUKO, the unit of effort has been the family farm and the family as work force, with the municipality and other institutions as backup where convenient. With the agrarian reforms and counter-reforms of the 1990s, the family is increasingly the unit of choice, undermined only by its small size. Even the research organized at the agricultural experiment station at Batallas is focussed on families insofar as it concerns socio-economic variables (COSUDE 2000A). This has not meant total disregard for traditional communal work, and there is still the use of *ayni*, sharing of major tasks between friends and relatives, and of *minka*, where a group combines but some direct recompense is given in goods such as farm produce, meals and drinks. But in both countries, the archaeologists' dream of totally traditional methods has had to be dropped.

Observing all these projects, both Bolivian and Peruvian, from the outside, it is apparent that there are conflicting interests at work. The projects are promoted by agencies, by authorities, and by individuals, whose career or whose *raison d'etre* is the project and who constantly present the technology in a favourable light. Their aim is to achieve maximum take up of the programme, usually measured as number of hectares completed. On the other hand, the communities, the families, even the individuals who are the objective of the projects, have very different ambitions. From conversations with them in virtually all the sites I have visited, their aim is survival, and they do desire projects, but primarily to provide some cash income or some kind of help. This is still the case, in spite of the fact (which I explained to many) that any kind of financial assistance had been dropped by all the agencies some years ago. This kind of situation does not seem to correspond to what has been attempted by the agencies, the adoption and assimilation of the technology for its own sake, over a period of time.

The Work of Reconstruction

The work of reconstruction has been done in phases. Surveying of the area to be used, its soil types and topographic slopes and relations to water sources, and measurements to define the areas under field platforms, and those occupied by canals, is a first job. The diagram, used in Bolivia, shows typical dimensions and the structure of the raised fields or *suka kollus* (Figure 4.5). This figure illustrates the need for careful construction of the platform edge with heavy grass sods cut out of the plain, to prevent erosion of this edge. It also illustrates the desirable content of the platform, with good draining soils in the centre and fertile organic topsoil piled in a slightly convex form on top. The ratio between the canal area and the field area is important, and it has varied from 30 per cent canal against 70 per cent fields, to 60 per cent canal and 40 per cent fields. Too large a canal area means a loss of cultivable land, while too small means a loss of the thermal protection provided by the water in the canals. Platform width is also important; too wide a platform means conditions approach those of the open pampa, as the tempering effect of the canals with their water is lost. A width between 4 and 7 metres is recommended, depending on availability of materials and severity of frosts encountered. In Bolivia, most of the projects which I saw were in fact outside these recommended limits, with broad platforms of 5-6 metres and only narrow canals between. For the Bolivians the possibility of mechanization and tractor use seems to have weighed in this decision. In Peru, narrow platforms of three metres or so were found in many early projects, where the instigators were keen to use traditional hand tools and no machinery.

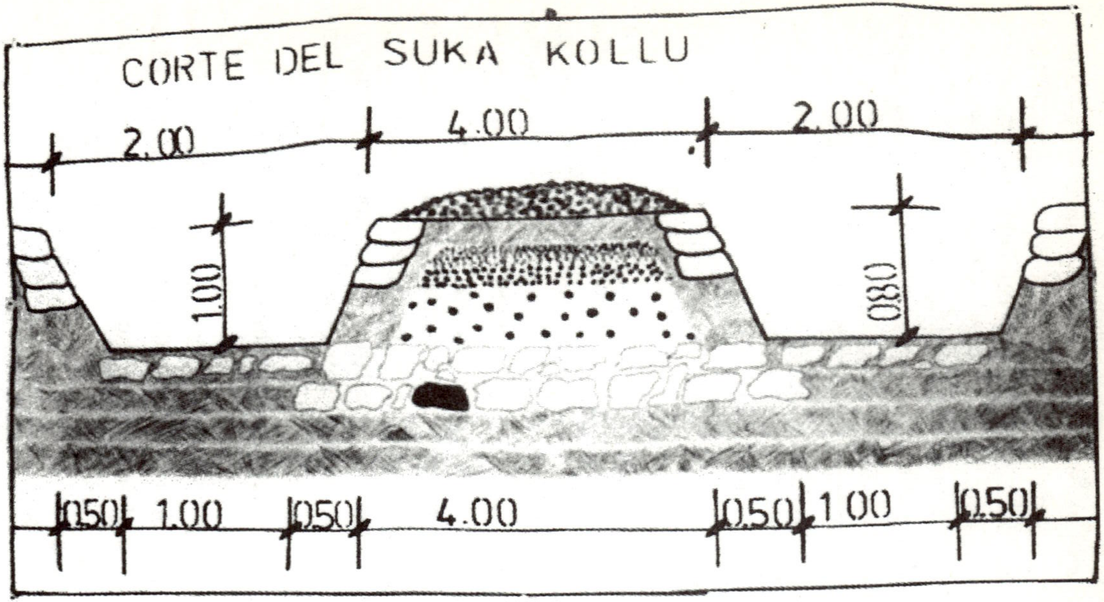

Figure 4.5 **The construction of raised fields. A diagram used in the field to help farmers shows dimensions in metres, and materials, including the reinforcement of field edges with cut turfs.**

Only from the 1990s was the move to broader platforms made, recognizing that a totally traditional approach was too idealistic.

The areal dimensions of each individual platform, and their orientation in relation to sun and slope, are mostly given by the visible remains of prehistoric raised fields. In some places, fields have been built without previous markers, and on steeper slopes, or with greater lengths, than the originals. But there are no hard and fast rules about what is desirable. A determining factor can be the source of water; in some places, where drainage is the main aim, the orientation has to be leading downslope so as to evacuate water, and in the growing season there is groundwater in the canals. In other cases on the higher parts of the lakeside land, the only source of water is rainwater, and here the orientation is made to conserve and keep in water, with ridges running across the slopes, or varying from one set of platforms to another. Where there is a river or stream supply of water to the fields, this determines the layout to some extent.

The boundaries between platform and canals have to be marked with string, then sods of topsoil cut, in the older projects with the Andean footplough or *chakitaclla*, to build first the retaining walls of the platform edges, and then to fill in the platform area itself. The height of the platforms is also variable but a recommended height is 0.80-1.00 metres as shown in the diagram. Where the topsoil in the area is thin, the platforms tend to be lower and the canals wider, in order to avoid filling the platform with infertile subsoil. This thinness also militates against the use of machinery, which has been tried in some projects and found to dig up subsoil and spoil the final results.

The best time for construction of the raised fields is thought to be in March-June, when the soil is still moist and after the harvest period. However, this period is often not suitable for the local labour force, and the work is deferred to later in the dry cooler season, when the soil is baked into hard clods. Then another traditional tool, the *waqtana*, must be used to break up the clods of earth as they are lifted.

The Direct Critique: Factors in Success and Failure

Physical Factors

For the most part, the raised fields have been well located in physical terms. The choice of location has commonly been governed by the pre-existence of traces in the ground of the ancient fields, and the old lines followed in general terms in the course of new construction. I observed only a few locations chosen where there had been an obvious and disastrous mistake – one was a river bank location where the flooding was excessive and the crops would fail because the platforms for construction were unable to rise above the floods, on the river Illpa near Juliaca. Another was an exceptionally dry site, where rainfall might have been expected to provide water in the ditches but ran away because of a highly permeable soil, just north of Puno. A third was a lake side site, near Pilcuyo south of Ilave, where the raised fields had been constructed during an exceptionally dry year when the lake level was at its lowest, and the land had been flooded ever since. In other cases, physical failure was due to regional effects like a general incidence of heavy frost, of flooding, or of drought, and no site could escape these climatic factors. In general, the physical functioning of the raised fields has been well understood. Other failures have related to the management of the physical resource, such as clumsy ploughing and digging out of ditches so that subsoil has been raised onto the platforms, or the failure to bring a stream in the neighbourhood to the raised fields so that they never had enough water. At Paucarcolla, detailed in the next chapter, there was flooding and crops became impossible because a road had been built across the natural drainage, damming it, a few years after the project was started. None of these problems were insuperable.

Other kinds of physical failure are partial and understandable. A main problem is the availability of the best quality of cultivable land. PIWA identified three types of land for agricultural purposes, *ladera*, at the junction of plain and hill, the best land for cultivation; lakeside land, with excellent soil and climate but special flooding and salinization problems; and *pampa*, the open plains, the least desirable land from most points of view. The *laderas* are already usually under cultivation by small farmers, as is much of the lakeside land. Agencies have been understandably reluctant to take out this productive land and ask farmers to convert it to raised fields. Land with a good supply of stream water is similarly often already in use. Thus the programmes, especially in Peru, have focussed on the relatively poor lands of the open pampa. 79 per cent of the land thought by PIWA to be priority land for programmes was pampa land. On these flat, open plains, the climate is harsher, the soils poorer, the liability to flood, salt and clay accumulations greater. Of the 60,000 hectares found by PIWA to be priority for their programme, some 43,000 ha. are dependent on rainwater only, having no river or stream to augment the supply (Zeballos 1994). This is a critical factor in a region with a rainfall deficiency.

Situational Factors: Ownership of Land

A first and perhaps most important factor has been land tenure. Both Bolivia and Peru have seen decades of turmoil and change in this area, as these countries have moved away from a semi-colonial situation of great estates with workers in conditions of near serfdom, in the 1950s, towards a more modern structure. Bolivia's National Revolution of 1952 led to an agrarian reform in the following year, to give land to the workers, and both communal ownership and private ownership of small plots were encouraged. In following decades, the communal reinstatement ran into huge difficulties. Partly because of the reorganisation of land, actual production declined and the farmers were impoverished even more, so that many left farms for the town. There had never been enough land to distribute to all farm workers, and most had to leave dissatisfied. Communal ownership itself was threatened by the disturbance that this caused, and now most land is run individually, although in many cases legal title to the land has not been granted.

A similar history obtains for Peru, where there were disturbances in the 1950s but a reform was only undertaken from 1968 by a military government which attempted again to reintroduce the ancient communities as a large scale economic unit. This again led to declines in production and disturbances as many workers found they were excluded from the benefits of reform and moved out to Lima or other cities. By the 1990s, most farming had moved into private hands, although the communities were still being registered as owners in the 1980s. Today, only remnants of land in small, often unwanted patches, lie in the hands of communities. But still, legal title to land has not been obtained by most farmers, and a situation of uncertainty haunts the farming sector as a result (Morris 2001).

In both countries the situation is unhelpful for small farmers contemplating investment in a major project. They may be uncertain as to ownership of the land and thus unwilling to even start a project. Or where communal efforts have been tried, the new structure of private ownership works against the idea of combining again to work common lands through joint efforts. The small farmers do not want, or are unable, to reinvent the past with their bitter memories of conflicts.

Economic Benefits

There are many studies of the economic value of reinstating raised fields. They commonly suffer from incompleteness, but they do at least show why the whole idea was so heartily endorsed in the early years. Many early studies simply looked at the yield levels on the raised fields from one or two years and compared these with those for the open pampa for the same time period. Others added in the cost of building the fields, but with the very low levels of payment for communal work, which, as we have noted, was difficult to organize and in the process of final decay. In Berastain (1992), for example, the costs of production are reckoned to be the same for both areas with raised fields and without, because the greater production from any given area of raised fields means a smaller investment in work per ton of production. His calculation (1992, p. 98) is that the yield on the open pampa is five metric tons per hectare, and on raised fields 15 tons. Since half the raised field is in ditches rather than growing platforms, the raised field yield per hectare has to be halved; this still leaves a net production of 7.5 tons per hectare, and this is 50 per cent higher than on the pampa. Since the costs are equal on both areas, the raised fields have 50 per cent higher profitability.

A later evaluation is that of Chire and Valdivia (2000), and it gives a more equivocal result, though the methods are still not wholly reliable. These writers use the full set of data from eleven different communities working with PIWA, and calculate the yields and profitability of the raised fields from the projects in each, comparing one community with another. The fields are found to be profitable in all areas, but only in three are they more profitable than normal farming of the same crops. In two years out of the six studied, the results are better, generally the wet years. But in other years, there is no difference detectable. These results could look worse if other points are noted however. First, the value of construction of the raised fields is depreciated over eight years. This might seem reasonable, but in reality, the great majority of projects I have seen are abandoned after four years. Thus the construction should be depreciated over that time period. Secondly, the yields and profitability are reckoned over the average of individual years' results; but these have not come from constant plots of land. Instead, the fields used in the community do change over time – a sector is abandoned, another is taken up as farmers try new areas and leave some of the older ones. Since the yields are known to be higher in the first year or two after formation, a falsely high average yield is obtained. If we were to take an individual piece of land, the yields must be lowered substantially over four years, not to mention what would happen over eight years. These adjustments bring the economic viability of the raised fields into question, and in each case would make them less profitable than normal cultivation of the pampa.

Yet another study (Izquierdo Condori 1992) uses some highly restrictive conditions in his assessment, and makes the whole enterprise still more dubious. His conclusions are that the whole technology is non-viable for the small farmer, because any investment must be recovered in one year, and this is the maximum amount of time a farmer, living always on the border of disaster, can wait. This means that the costs of reconstruction of raised fields must be covered by the first year's yield, an impossible target. For commercial large scale farms engaged in arable cultivation it would be different, but such farms are not found in the region, and they were never the target. This kind of assessment seems too severe, and I did not meet anyone who thought the returns on labour and time invested should be made in full within one year.

A Different Economic Analysis

In reality, no full cost-benefit analysis of the western type can be applied here. Unlike a fully capitalist economy, where all factors are employed fully, farmers here are employed only part of the year, and while some move out to earn a little money in the low season after the harvest and before ploughing, in the period June-September, or again after planting and first cultivation but before harvesting, in the period January-March, many are idle, and their time and labour has little cost.

For these people, the cost of building the raised fields could be slight, and their economics might be compared to those of Chayanov's peasant economy, observed in Russia in the early twentieth century (Thorner *et al.*, 1966). Chayanov saw the resistance to mechanization and modernization in old Russia and found logical answers to his questions in the management of time. In the Russian and Ukrainian plains, lands of dominant grain cultivation, the soil is prepared, the crop is planted, tilled and harvested, but there is then a long period of semi-idleness. For these people, it did not make sense to employ tractors or other machines to speed their work, as they would have made a major investment with the result of having more idle time. They would rather be busy with mending their wooden hand tools, looking after animals, and making clothes or repairing their wooden houses for the Russian winter. Andean small farmers are not in quite the same position; their farming is more diversified, some now have alternative work in the nearby cities. But in any case, the economics of capitalist farming, with the aim of reducing labour costs to the minimum, do not apply.

The time period for assessment is also critical. There are no cases of continuous use of the raised fields for eight, ten, or 15 years, which may well be more relevant time periods over which costs and benefits should have been measured. The relevant time period has in fact not been established, and it would depend on the level of fertilization possible, and the time period for recuperation of soil fertility. This makes the whole economic benefit in normal terms, subject to considerable doubts. All the small farmers I spoke to considered that the fields needed to rest after two, three or four years, and could not say when would be an appropriate time for ending the fallow, though it had to be at least seven years. They pointed to the high terraces, where twenty years might pass between periods of cultivation. One significant positive point however, and which will be returned to again, is that raised fields have produced crops when nearby normal fields have failed. This was notably the case over the region in 1985-86, when crops failed through floods on the pampa, and locally in many places since, when frost has struck normal crops. In sum, there is no question of working out the cost or the benefits of individual years or even an average return, but having a resource which at some time may be called on to provide protection against starvation or economic disaster.

Sustainability

A different kind of economics comes into play when the emphasis turns from maximizing productivity, to ensuring the maintenance of the resource over the long term, and this is the idea that lies in the name sustainability. Beyond the concern of farmers and agencies, for the very survival of rural communities, there is a concern for long term sustainable development, which is understood on several levels. From conversations with Alipio Canahua, CARE project director in Puno, he sees three elements to sustainability: physical maintenance of the land resource through intensive use of the raised fields, combined with rest and restoration of the slope lands traditionally used for crops; secondly, economic sustainability, providing money incomes which allow farmers to rely solely on their altiplano farms for support; and thirdly, a cultural sustainability, since Andean technology is being used rather than foreign capital based mechanization. These dimensions are reinforced in organization terms, by attempts to strengthen linkages and functionality of different levels of grouping; the family, the community, the production unit, and second level organizations which bring communities together.

The raised field projects certainly do promise something far more sustainable, in a poor region of a poor country, than other projects for the highlands. To give a couple of negative examples, in Peru, there have been large funds put into a number of projects for rural development of the altiplano, including much mechanization and capital expenditure on infrastructure. One of these, begun in 1982, was a scheme for erecting a 4.5 metre dam on the Rio Illpa and irrigating up to 6000 hectares near the Illpa Agricultural Experimental Station, halfway between Juliaca and Puno (Erickson and Candler 1989). This never flourished in the 1980s because of land tenure disputes, and only now is being finished partially, with under 500 hectares of land being irrigated. Many of the tertiary canals to feed into farms have not been built, though the main cement canals have been constructed. In probability it will never be successful, because the waters of Lake Umayo, which feeds the Illpa, are saline and salt accumulation would be a major problem with long term irrigation (Zeballos 1992, p. 132).

Another example is the development by the National Agrarian University, based near Lima, of large-scale raised fields in the form of wide platforms. These could only be built by heavy tractors that would seriously damage the fertile topsoils of the altiplano, at great cost, and again would replace the prehistoric remains of fields constructed with the same purpose and built with simple hand labour. This project has fortunately been abandoned.

We shall return to sustainability of the technology in Chapter 7. Any kind of production which aims to be sustainable in the Andean Highlands must be well integrated with the rest of the farming system; the converse is probably also true, that no innovations will be allowed to come in which do not contribute towards a sustainable farming system. As suggested also above, sustainability can be taken in cultural terms too, so that the innovation needs to be accepted by Andean communities, groups of people who are not necessarily looking for traditional ways of working, but are most welcoming to ideas which they can bring in themselves and incorporate without changing their lifestyle.

Social Elements

While economic analyses of various kinds have been conducted, study of social issues is limited. The only extensive socio-economic study made (Salas *et al.* 2000) looked at a widely distributed sample of 204 families. This focussed on the differential impact of the programme over the period 1995-98. They found, at geographical level, that the cultivated land areas away from the lake, what is termed the Suni zone, roughly from 3830-4000 metres, had the largest areas of raised fields, but that this simply reflected their larger farm sizes. For all areas, lakeside and Suni, the better-off families had done most with the raised fields. For all areas too, the younger farmers had done most with the programme, and age was a problem for participation. Concerning the effects of the programme on the levels of living, they found positive results for all. Table 4.1 illustrates their findings in terms of value of farm production.

Table 4.1 Value of family agricultural (non-livestock) production in 1995 & 1999 (in constant soles 1995)

Area	Before	After	Change
Lakeside			
users of *waru waru*	4294	7509	+75%
non users	2731	4533	+66%
Suni			
users of *waru waru*	3444	6238	+81%
non users	1275	1615	+27%

Source: Salas et al. 2000

From this table, a huge apparent increase in income is available to users of the technology, although non-users also managed to increase their incomes dramatically. But of course the figures derive from measurement made during the actual course of a project, and do not measure the later state of affairs. I was unable to find comparable measurements for communities after the projects had ended. The greatest effects were on the *suni* areas, which is surprising because the technology has been most applied in areas near to the lake, but the greater poverty of the interior areas is no doubt important in this finding. For the poorest communities, the presence of a major project is of great significance, especially in remote areas where no other work is available for much of the year.

This study and others like it point out the high participation rates in the projects. According to those interviewed, the choice of site and the initiative to actually start construction were taken, not by the agency but by leaders of the community or the individual farmers involved. Participation in these activities is reckoned normally to be a key variable in deciding the sustainability of a project. Only if the local people themselves have made decisions and taken part in all aspects of the project, can long term survival of the innovation be expected. But as the next chapter will show, even high participation rates are not enough.

It is well known that women also took an important part in the raised fields projects both in Bolivia and Peru (Quispe, Diaz and Peralta 1999, PROSUKO 1995), and women-led projects have been some of the most successful ones on the altiplano. In some cases, it was because of the existence of a strong social institution, the *Club de Madres* or Mothers Club, that a project was possible. In such cases, the Mothers Clubs were the strongest institution in the community, and seem to have had special importance in areas where the men were away, migrating out for seasonal work on the coast, because they, the women, were in the community permanently and able to manage the project. In other cases, where a whole community or a family adopted the project, the women had a subordinate role in the decision to adopt the technology, and in the hard work of actual reconstruction, they were very active in the use of the fields. Women were in charge of choosing seed, fertilizer spreading, planting seeds, and selection of the product for sale or consumption; men were dominant only in the digging of the land, or its ploughing with the yoke of oxen, and, curiously enough, the use of sprayers to control weeds and pests. Presumably this latter specialization was due to the considerable weight of a spraying knapsack when full of liquid.

The Special Context for the 1980s Success

If the social and economic results are so limited, when measured in standard western form, the question arises how the programmes for reintroduction ever started or achieved such enthusiastic reception and support from institutions in the early years, and a partial explanation may be offered here.

Many of the projects achieved rapid take-off from their start in 1986-87, because of the flooding that had occurred in the previous season, 1985-86. In that year Lake Titicaca had overtopped its banks, various tributary rivers had overflowed, and rainwater had also caused flooding on flat lands near the lake. This flood year had been preceded in 1982-83 by a growing season with severe drought, and in 1983-84 by frequent frosts, which made many farmers already dependent on aid from government agencies, so that the raised field projects were an ideal mode for providing work to rural people, with the possibility of later benefits from crop harvests. Payments became a norm because of the emergency situation with the flood following the drought; this emergency meant that the Peruvian government had already set up a system of payments for farm workers and special projects, so that when the raised fields projects came in, they had to compete with the central government schemes. Once the payments had been established, and after a few years of the system, it was difficult to stop what had become an expected feature of the projects. The agencies were aware of the problems of continued payment, the dependency on aid which is known as *asistencialismo*, but were only able to stop it from about 1996. As noted above, a new project or rather the possibility of a new project, is still today seen as a promise of wage payments or some kind of money income, in a rural area of grinding poverty. There are thus at least two differing discourses about the projects and their meaning for farming in this region.

In many cases, as in the area around Huatta, there was also a convenient situation in land tenure changes which helped the projects. The breakup of the great haciendas such as Buenavista, adjacent to Huatta, meant that many peasant farmers were left without land, or with land far from home, or without any money income. Projects could supply work and money income, and they held out the additional promise of production of useful crops for subsistence. Buenavista had long ceased to be a traditional hacienda of the altiplano, following agrarian reform under the military government in 1969 (Morris 2001). But its basic organization had not changed, as it became a SAIS, a large scale cooperative where the state gave the lead, where there was a central paid work force, and where extra labour was gained from the villages around the estate, in return for some payment and a small part of the profits. This structure only broke up in the 1980s with further reforms causing a total redistribution of land of the estate, and chaotic conditions for many farm workers in the villages.

Another condition which might be construed as favouring the programmes for the raised fields, at least in Peru, was the gradual but continuously increasing pressure on the land. Population in the rural area was on the increase in Puno province between 1950 and 1990, in a period before much rural-urban migration had taken place, but when modern medicine was beginning to reduce infant and general mortality in the countryside. Many villages suffered an increase of 30 per cent in their population over this period. In the 1990s, the annual increase in Puno province continued, at about 1.2 per cent overall, and 1.0 per cent in the rural area (Avila 2002). This population pressure translates into immediate pressure on the land resource, as the rural population is essentially dependent on farming for livelihoods. But demographic pressure cannot be used as a major positive factor on its own; the same villages which have had a massive increase in population have mostly abandoned the *waru waru* after their projects ended. In Bolivia, in any case, the population has been declining in the same period. The area of PROSUKO's work on the altiplano suffered a decline from 215,000 in 1976, to 188,000 in 1992 (PROSUKO 1995). This latter region has suffered enormous conflicts over land in recent decades, and has the urban attraction of the capital city, La Paz, which has syphoned off much of the young population.

At the same time, attempts at increasing the intensity of production on the land were producing problems, in erosion of the soil and loss of fertility. Today, and for many centuries past, it has been the practice to grow crops only on the slope lands immediately above the altiplano, as the still night air and powerful outwards night-time radiation causes frost and cold air to accumulate on the flat pampas of the altiplano. But the slope lands are also those most susceptible to soil erosion, so that the system of slope cultivation has been gradually falling into crisis, and extra lands are urgently needed. Achata and Sanchez (1990) cite the example of the community of Anccaca, 12 kilometres west of Puno, on high land with a more severe climate than the immediate lakeside. Its 400 plus inhabitants have access to less than 200 hectares of slope land, insufficient for the production of enough crops of potato and quinua in normal years, and far from enough in poor seasons. Here, the project of planting the open pampas with crops on the raised fields was taken up with great enthusiasm, not simply because of its technical possibilities, but evidently because of the urgent need for subsistence crop land to replace the worn out slopes of traditional use. This project was run by the local agricultural research centre and another development NGO, supported by Canadian funds. The project was successful for several years, but is now unused, raising many questions about viability over the long term. We shall return to this question again in the next chapter.

Broadening the Programme

Throughout my observations of the programmes in Bolivia and Peru, I was struck by the limitations of the projects due to the narrow focus on the raised fields. Because of the enormous success and perhaps the academic prestige of the first projects, whole programmes were set up to establish raised fields without any other objective. Ironically, given the non-Green Revolution credentials of the programme, this was a kind of Green Revolution, depending on a technical breakthrough and nothing else. This problem was increasingly realized by the agencies, but to change their orientation took time.

In several cases there have been shifts in the organization of the raised fields projects. The moves in both Bolivia and Peru from communal to individual farms and farmers has been mentioned as one move away from the emphasis on traditional structures. Another general change was the move away from providing any money incentives. There has been a move away from a totally organic farming prescription, especially in Bolivia, to the use of chemicals and where available, machinery such as tractors. Chemical pesticides have been seen (in use with knapsack sprayers) also in Peru, though there is a more general effort to use organic methods. Tractor use is certainly frowned on in Peru, and this is possibly a major drawback to the future of the programme. It is true that digging out topsoil with large tractors for constructing the raised fields is fraught with danger, principally with the danger of digging up subsoil which lies often only half a metre below the surface. But this should not have been an insuperable problem. Another matter is the use of wide cultivation platforms, up to six metres in width, as seen near Batallas and noted in COSUDE 2001, to make tractor use easier. This makes mechanization easier, but may risk losing microclimatic protection, especially in Bolivia where air and soil humidity are lower and frosts become progressively more acute.

But these are internal adjustments to the programmes. A more general broadening has been seen as necessary by some agencies. After the initial enthusiasm period, the failure of the raised fields to link to any other element of the farming system became a real obstacle. The PIWA organization in Peru was hampered in trying to make changes because its remit had been set as reinstating the raised fields, but the CARE programme did expand out vertically from about 1995, into storing and marketing, and into a number of minor social activities. In Bolivia, the orientation was changed, at least in intent (COSUDE 1998) by adding on to the raised field projects some aid for farmers in other activities, notably crop cultivation on the open pampa without the benefit of the raised fields. The purpose of this new line is unsure; perhaps it is an attempt to show farmers that raised fields will give better crop yields than on the open plateau lands.

In general, despite these efforts at broadening, the projects do not seem to have been integrated with the general farming system, and raised fields are treated separately, as the special exceptions to the rule that they still are. This is a key matter, and it will be explored in Chapter 7. In some places, the situation is even worse than lack of breadth, in that raised fields are seen as land uses for marginal areas only. In Peru, as has already been discussed, the decision was made to leave out of the priority areas much of the best land because it was already in intensive traditional cultivation, and to attempt to colonise the lands where little present cultivation was competing. In Bolivia, the raised fields have very often been used to occupy patches which suffer from poor drainage (in Aymara the *poruma* and *joqho* lands) lands which are not normally used at all or which only occasionally have an agricultural use when there is an exceptional drought (Rocha et al. 2000). In interviews conducted as part of a social survey of the technology in the 1999-2000 season, the preference of farmers in choosing land for raised fields was seen to be for the *karañas*, the seasonally flooded lands, or for *poruma* land, with soil of good quality but subject to flooding (COSUDE 2001). These are patches of land which are normally never brought into cultivation at all except in very dry years. The importance of this marginalization in areas and within the farming system is that Andean farming is not highly specialized but based on the integration of many elements, across time and space dimensions. One whole area of study in the universities of the region has indeed become Andean farm technology, and it is far wider than raised fields.

Differences in vision are seen not only in reference to the areas to be used and the function in the system, but as to what projects are supposed to bring to the farmers. There are separate discourses, different sets of messages, coming from different actors in this vast play of technological change. On the one hand, the presenters of the technology are mostly the agents who come from the city and wish to bring the new idea to the countryside. Their presentation is nearly always positive, though in some cases they have rejected initial advances from communities because their sites are too difficult physically or too remote. On the other, the discourse from the objectives, the rural communities and their component families, is one, not of rejection, but of will for more projects, more help, because of the financial support it will bring. Perhaps the whole programme went on too long, because there was agreement on the need for projects, but no congruency on the reasons for such projects. There was of course another set of actors, the early archaeologists and other scientists, who also wished to promote the technology as it helped or suited their own academic task. They in general, were even more positive and optimistic than the agencies, and gave these latter much support in their initial takeoff. Was this an unholy alliance?

5 The Situation Today: Some Case Studies

Introduction

It is not possible to make a strong positive assessment of the programmes to date, and a first judgement would be that there has been a failure to make the transition from supported project to incorporation of the technology into ordinary farming. Both in Peru and in Bolivia, the early projects are all without activity, and the only working raised fields are those in current or recent projects, on the rural fringe. This is true in all kinds of site, independently of physical site, remoteness or proximity to towns, for both community and family based projects, and for different kinds of promoting institution. Certainly, this was not what was expected from the great programmes begun and carried through by the agencies. Nor is it the picture presented through published results, where any visiting scholar can read of general enthusiasm, acceptance, maintenance of existing projects, and success in changing the level of food autonomy for small communities. Perhaps a more realistic assessment is the decision, on the part of the agencies, to terminate the programmes in 2002, with the end of the *waru waru* projects of PIWA, an agency which has moved on to other programmes, and the finalization of the programmes by CARE, also in Peru. On the Bolivian side, there is continuing activity, but at a low level, and concentrated in just a few small pockets where relative success was found. Certainly there is no general extension over the whole of the altiplano.

But there are some good results, and a structure which seems to be working. The process is best illustrated by a set of examples, both negative and positive. Those chosen are simply ones from which I was able to gather more information, by visits over a period of several years, and by additional documentation from the agencies. I spoke to country folk, to the agents managing the projects, and to some of the community leaders in each case.

Life Today in the Countryside

Visits to communities which have benefited from the agencies and which are anxious to please visiting scholars or extension agents can give a feeling of positive attitudes and hope among the small farmer population. But this can be very misleading, for the demographic and economic situation is difficult, even for those families with clear tenure of sufficient land, and with access to work outside the farm. One striking statistic is the pressure of population. In the Bolivian sector of the lakeside, 45 per cent of the rural population is in the age group five to 24, meaning that there is a huge increase in young people who will need feeding and clothing over the coming decades (COSUDE 2000). Against this statistic, the population is actually decreasing in the rural areas, so that the massive increases are being syphoned off entirely into the cities, and the great majority into the city of La Paz, presumably because the farming populations cannot support more people. In the rural population the average household has between five and seven members. The average income of these families was, in the year 2000, between 3800 and 4900 pesos bolivianos, or some £400-500. This money will go further than in Europe, but still, a hotel in La Paz may cost £50-£100 a night. Evidently, this is a subsistence economy, with no room for manoeuvre or interest in work which is not guaranteed to be productive.

Most of the children in the families can be seen in the countryside as few finish their schooling. PROSUKO's survey data (COSUDE 2000) indicate that 35 per cent of primary school age children do not attend, and 50 per cent of those at secondary level. These children are employed on the farm and help their parents plant crops or look after the animals. Among the adults, many migrate out to the coast after the harvests between June and September, to return when sowing the next year's crop is planned. In Peru, an estimate from one small area (Schulte 1996) is that 35 per cent of the men migrate out temporarily for work on the coast, mostly in the sugar and cotton plantations.

It is also worth insisting that the economy is still a subsistence one in many ways, and that it is unlike North America, Europe, or even the larger part of Latin America in the way that labour is used. At harvest time, it is possible to see men, women and children out cutting the crops of barley, oats, or wheat with a small sickle, then stacking the crop in stooks to dry. After the harvest in May, the men may be seen beating the crop with wooden batons or flails to separate out the seed, and then throwing the crop in the air to winnow it, on simple, dried and cleaned patches of earth. And the resulting grain harvest is carried away on men's backs to the stores. At planting time, it is again a case of manual labour, often by women alone, of individual potatoes or casting of seed. Throughout the whole set of operations, there is frequently no use of animal traction or of machines. This is a situation more comparable to India or China, than of Latin America. The most common use of animals is in ploughing, when a yoke of oxen may be employed to turn over the heavy soils. Even on this point, many small farmers cannot afford to hire oxen power and use their own strength and the Andean digging spade, the *chakitaclla*, to turn the soil.

Peruvian Raised Fields

The Example of Caritamaya

By far the larger area of experimentation, and the greater effort by agencies in modern time, has been on the Peruvian side of the border, and we will start with this. In recent years, the most successful raised field experiments have been under CARE International, whose programme in Peru has only recently terminated, and where ongoing raised field use is in evidence.

For CARE, one of their best examples of success in taking projects forwards over a period of time, has been the development at Caritamaya. I first visited this settlement, in reality a group of five communities, in January 1999. It lies south of Puno, eight miles southeast of the town of Acora. The land is fertile, with abundance of water, and with relatively good access to Acora and Puno. The land was formerly in great ranches, farmed principally for sheep with some cattle, at a very low level of intensity, perhaps two sheep per hectare. Now because of the pressure of population on a limited land resource, there is a context where sustainable but intensive cropping is needed to support the rising population. In bad years, fifty per cent of the adult male population has to go out, mostly to the coast or to the cities, to work for a money income, though this figure has declined somewhat in recent years. In this area, traces of ancient *qochas* and *waru waru* were found, and formed the basis as elsewhere for rebuilding both these features. In 1999 there was evidence that they were in active use for a rotation of potatoes, quinua, barley, and alfalfa. Today, the most significant areas of restoration of the raised fields are often associated with the depressions where there were once *qochas*, or at least where there is poor drainage. This feature seems to be a key to adoption of the technology today.

Anyone visiting the area has first to go through the town of Acora, and this is the local market centre. It merits attention, because this is where the drive force for any farming programme in the surrounding region ought to lie. Today, and probably since prehistoric times, it acts as a fascinating place of trade, an interface for exchange between the mountains to the west and the plains around the lake. The hill people bring to the open air market live cattle for fattening, pelts, salt and cheeses. The lakeside people bring onions, grains of various kinds and potatoes, which they often exchange directly for mountain products, especially the poor who have no money at all. It is thus a barter market. There are also many stands for townsfolk, selling soap, buckets, second hand bicycle parts, cheap clothes. There are also the vendors of services, selling incongruously side by side, haircuts, fortune telling, funeral services, and pregnancy testing. A couple of converted houses on the square have machines set up in them to offer the milling of barley and oats, for small farmers who will take the product back for consumption at home. But this is a market of the poor, not generally offering products to the national or even the regional markets. There are few surpluses to sell. One small exception is the wool and fresh pelts from the mountains, which are sold here, taken to Juliaca by merchants who bulk them up for transfer to the woollen mills in Lima or Arequipa. For maximum success of projects such as the raised fields, market surpluses would be needed and means of marketing them efficiently. These structures are not in place today. The quick answer might be that a successful project might change the marketing and modernize it, but the investments need to be large and this place will require much time and investment to make it a dynamo for the regional economy.

Passing through Acora and going out to the hamlets nearby, one can see most of life in the open, flat countryside. On one visit with Ernesto Chura, in the space of a few hours, I visited the village of Thunco, where most of the locals were attending a meeting for sheep washing, as a local vet had offered to coordinate this exercise and provide his skills. From there I went on to a separate community at Ccacca, where ten men were hard at work treading mud with their feet and adding water, then stalks of dry grass, to make adobe blocks for a new house. Another few miles from here, at Culta, another group of people were encountered in the farmyard behind a house where a neighbour had died the previous night. His family, but also the whole of the local community, had come to pay their last respects. All these communities had had experience of the raised fields, they could show the sites where the technology had been practised, but they did not now cultivate them.

There is written evidence to show how the raised field programmes came to Caritamaya, and in particular, how local organization functioned. In 1979, students at the university in Puno with local roots, at a time of considerable instability in the countryside, led a movement to take over some of the lands of an *hacienda*, the Fundo Chillerota, bringing together five local communities (whose homes and land was part of the *hacienda*) for this operation. This successful takeover meant that the five communities, including Caritamaya, were able to build up an administrative structure and maintain some solidarity with each other over time. Each community was composed of under four hundred people, and it was relatively easy for a simple leadership organization to keep them together. The example given by Vicente Ventura is when in February 1980 they had to work communally to light night fires to keep off the heavy night frosts at the time of most danger for the potato crop, when the young leaves and flowers are most susceptible. But over the years, with the process of parcelling out the land so that no communal land remained, the level of cooperation declined and resentment grew up because it was felt that the leaders of the community had been given the largest and best plots of land, because of failure to turn up for cooperative work of any kind.

In this area, the farmland per family is only one to two hectares (Ventura Lope 1992). The cultivated land is mostly worked in *aynokas*, i.e. in open fields where each family has their own strip or plot. Vicente Ventura fears the time when each family will have only one or two *surcos*, or strips (roughly 1/10th of a hectare) each, as land is constantly subdivided between children in the large rural families.

Caritamaya, according to Ventura Lope, made an early start to the raised fields. Initially the push for organized work came from the bringing together of the local farmers to work on projects for the PAIT, the Programme for Temporary Income Support, set up by the central government in response to the drought of 1983-84. Work was first done on raised fields as part of this programme. The central government offered wages for the work force, a minimum salary for up to sixty people. There were three agents for the programme working permanently in the area, which ensured progress, and the building of two hectares of well constructed, high raised fields. In the following years of frost (1984-85) and flood (1985-86) the raised fields performed well, and the local people took over and managed the project alone. PIWA entered the scene in 1989, and was also useful in promoting the projects already set up, with initially some 100-120 families involved. Apparently with this institution, the payments were lower, and many farmers left the work after one hectare was completed, leaving some 40-50 families. PIWA levels of compensation, here as elsewhere, were set at $340 per hectare, plus $100 worth of tools, and a loan of potato seed sufficient for 50 per cent of the area constructed. The yields obtained were high – up to ten tonnes per hectare. However, only a small area was actually involved – some four hectares of land. Taking out the canal surfaces, this translates to 2.5 hectares of actual producing surface. According to Ventura, the large number of people leaving was due to the toughness of the work itself as the soil was very hard and had to be broken up as construction went on. This work was done during the dry season, when the soil becomes bone hard, and obviously better results could have been obtained by working during the wet (growing) season, and by sacrificing some time in the farmers' individual plots. Following this, the next year the Caritamaya group, or those still interested in *waru waru*, decided to work from March, when the soil was moist, and made much better progress, constructing some seven hectares. Their agreement was to use one of their three main open fields (*aynoka)* for raised fields, and this was carried through in part. The *aynokas* here are 30-50 hectares each, and seven hectares were constructed.

CARE made contact later with this group in 1990, and its work has been more at a social and organizational level. Initially, the Caritamaya people had projects going with both CARE and PIWA, in the season 1991-92 producing on 13 hectares of raised fields, but PIWA dropped out of this situation of competition and CARE has continued. However, the raised fields themselves, from observation and commentary, have not really progressed. Whereas 120 people had worked the *waru waru* before according to Ventura Lope, now in the year 2000, there were only 32; farmers were spending more time on their private parcels than on the collectively managed raised fields. There is no communally owned land, which creates all sorts of problems and eventually a lack of commitment to the projects. What has succeeded with CARE, is the structure of administration. The emphasis has been on working with, not of the whole community, but of groups of a few, usually about ten, families from each of the settlements. A higher, second level of organization has been erected from 1997, the Asociación Rural de Productores Agropecuarios del Altiplano (ARPAA), with Vicente Ventura as its president, and this administration is used to coordinate efforts and attempt marketing of any surpluses. ARPAA helps the small farmers in various ways, notably in the complicated field of land tenure, where most farmers have problems about such matters as the legal titles to land, surveying of boundaries, and land transactions. Efforts are now being made to broaden the economic base of the communities, with stimulus being given to craft work such as the making of alpaca goods, and to improve soil and provide forage for animals by encouraging the planting of alfalfa. In animal husbandry, a bull is owned for servicing the few cows owned by many small farmers, and ARPAA has a veterinary office. Their principal need is seen as a tractor, but this lies beyond the financial capacity of the organization.

The existence of ARPAA has helped contact with other local organizations. One of the strongest has been the Mothers Club at Caritamaya; like other such clubs in the altiplano, it exists to help and organize the work and social life of rural women, who are often left with much of the farm work, when the menfolk have to leave to find seasonal paid work in the cities, or in plantations on the coast of Peru. CARE visits the Mothers Club here at irregular intervals, to hand out prizes for craft work done, to instruct on specific farming problems, and to hear of and advise on projects for the future.

Technically there were new experiments at Caritamaya, a place where success for the agencies in building raised fields gave them confidence and the will to look at another ancient form, the *qotas* or *qotañas*. Remnants of these pond features were found in the area, and they were re-established in a novel way, by combining their round form with the raised fields. Previously, Ventura says that there were attempts to cultivate these hollows, but usually this was fruitless because of flooding. With good drainage however, their rich soils provide a good harvest and in dry years, the contrast with surrounding fields is striking. Here, a design was brought in by engineers from PIWA, and followed by CARE. The results were good harvests of all the main subsistence crops, potatoes, quinua, cañihua, and barley.

None of the five communities had greater success in the long term, if we measure this by the permanence of cultivation, with either the *waru waru* or the *qotañas*. In October 2000, according to Vicente Ventura, only about 1 hectare in each of the communities was in use for *waru waru*, and none of the *qotañas*. Even the *waru waru* which had been built recently, in the late 1990s, were out of service in 2000. This was explained by Vicente as the result of having had a relatively good set of years climatically, and productive soil on the open pampa, meaning there was no need for the raised fields. Native weather forecasting was being used to predict the bad years and 2000-01 was going to be an average year. (This prediction in fact turned out to be disastrously wrong, with extensive flooding in the region, but it is recorded as part of the reasoning for low use.) In one matter he was insistent: there is no failure of the raised fields. Their technology is known and appreciated, and will be used whenever convenient. He reminded me that in the olden days, the land was not used continuously every year, but rested after a four year cycle, perhaps for seven to ten years. This has not been happening recently, but it should be the norm for maintaining the good heart in the soil.

It should be noted that there is no public admission of the poor results in terms of areas, on the part of any institution. CARE records that there are 55 hectares of *waru waru* in Caritamaya, but this is a total constructed since the 1980s, and the amount actually in use is one tenth of this, often on the poorly drained hollows which points to their use as a way to use otherwise unusable wetlands. In each of the five communities, I asked directly why the projects had been started and in each a major reason had been to use the *qotañas*, the round ponds, since they could not be used at all otherwise. Nonetheless, the example of Caritamaya must be regarded as a relative success, where the idea and the use of the raised fields has remained in force over a considerable period. Success must be imputed, partly to a dynamic and permanent leader in Vicente Ventura, partly to the poverty of the farmers and their urgent need for aid.

Paucarcolla

Something of a contrast is presented by the project at this settlement, some seven miles north of Puno a few hundred inhabitants on the open altiplano, but sited beside the main road to Juliaca and thus with good access to urban services. Although the houses sit either side of the tarmac road, this is a poor place, of one and two roomed houses in brown adobe or simple brick, with a few stores selling basic goods. Around the town there is open altiplano country, mostly grasslands with a few cattle, some of it under water in the last floods in early 2000. The local people regard the land as of good quality, but some of it shows signs of salinity, and the drainage is generally poor. The project reviewed here is a single effort, rather than the broad programme administered by Caritamaya, and only a few people have been involved, but it shows some similarities in the approaches to land use and similar problems in long term management, and some major differences in scope and scale. A road built by ranchers crosses the surface north of the town, damming back surface water and creating problems for the small farmers. As on some of the most successful sites in Caritamaya, it is on this poorly drained land that the raised fields were constructed, always a marginal site.

At one of the little stores, with an earth floor and with sacks of seed potatoes vying for space with soap and toilet rolls, I spoke to the owner's wife, Gladys Rodriguez Portugal, president of the *Club de Madres de los Milagros*, the local Mothers Club, a woman in her sixties and not wanting many new challenges. Like Caritamaya, the positive factors of great need and a strong leader have kept some kind of a structure in being. The name of Gladys' store is the "Bodeguita Señor de los Milagros", the Lord of Miracles Store, and it is certainly some kind of miracle which is needed if anything is to come of the raised fields here.

A major purpose of the call for a raised field project was the need to raise money among people with little or no regular employment. Some of the members were from totally indigent families, with no source of income at all, and the club was formed in 1990 partly to try to find some solution to their problems. The approach to PIWA was made by the group in that year, and agreement quickly reached after consultation and field visits from PIWA. This agency's offer was attractive because there was payment at that time for work done in reconstruction, for seeds and tools. The enthusiasm of beginning brought in 45 members. Now there is no help and there are only 15 members, though whether they will ever combine again is uncertain.

At Paucarcolla, the fields were built on the remains of prehistoric ones, and following the same lines, on land near the village. Alipio Canahua (now project director for CARE but then working for PIWA) advised on the project, and one hectare was built in 1990. This is a mothers' club, but there were some men members, and men and women worked together in the reconstruction. As is the norm with PIWA, the project lasted for four years, in terms of funding, and at the end most of the land was left to rest. The total of raised fields actually built was about 12 hectares, though at any one time the amount actually in use was never more than about 2.5 hectares.

Some of the members are still men, landless labourers whose situation is urgent and who are unable to move elsewhere to find work. Most of those participating have no land of their own apart from small garden plots. The organization of the work and sharing of the product during the time of the project, was on the basis of equality. Each member had to contribute labour, and each shared equally by weight in the harvest of crops, measured at Doña Gladys' store. What is evident at Paucarcolla is that there has been a drift away from the collective work on the raised fields, into individual small workings of the raised fields, plots to raise just a small amount of potatoes for the home. Doña Gladys admits this and explains it in terms of the lack of support from the agencies. PIWA would only give aid to collective action, and there is no support for individual efforts. This is unfair in her eyes. There is a long tradition of working as a community or as a smaller group, but also, every family has traditionally worked its own land as well.

In the season 2000-01 about one hectare was reworked, with no help from PIWA beyond consultations, the loan of potato seed to be paid back after the harvest with 30 per cent interest, in other words, 30 per cent more potatoes than loaned, and the promise of alfalfa seed for the period after use in annual crops. In the following year, 2001-02, nothing could be planted, as the previous crop had been flooded and there was nothing to give back to PIWA. In such conditions, the group has nothing in reserve, nothing to fall back on, and the whole programme, for the moment, has died.

As elsewhere, the project has had to contend with land tenure problems. From the beginning, it was possible to reconstruct the *waru waru* because they were on communal land, recognized as such from time immemorial. Once the land had been put to *waru waru*, there were attempts by livestock owners to claim it as their own. This situation was only resolved by the Muncipality, which declared the land as municipal in 1992. This declaration was in order to protect the land for the Mothers Club. But the land available is a small amount, and its quality is dubious.

From need, the Mothers Club has sought to diversify into other work. Some work was established in processing animal skins, tanning and leatherwork, but this work is at handicraft level, and cannot be efficient or fast. It would need capital to become properly established. There was a *huerto familiar*, a system of family gardens for vegetables, established in 1990 by the Micro-Region Puno, a central government agency for development, which helped people in the town to cultivate salad crops, beets and the like. Once started, the vegetable gardens are maintained by using sand, manure from the animals stabled nearby, and compost from waste heaps. Again, this project has had a sort of success in identifying other kinds of production and improving the diet of the people involved, but it is not a money spinner and only one or two small private gardens now remain.

There appears to be a problem of succession at Paucarcolla too; Gladys is seen by others as very important to the project, and though she wishes to retire, there are no natural successors. The Mothers Club has an important presence in a variety of communal activities – village parades, fiestas, sports events such as football, through the year, and all of these are attended by the club. The future here is obviously uncertain. If the leader goes, there is little chance of continuation. But a revival under present circumstances is in any case dubious, because there are no funds, only debts remaining from the great effort put in.

Paucarmayu

Further north from Paucarcolla, and west from the main road, Paucarmayu is a hamlet-sized loose settlement of dispersed adobe huts with tin roofs, formed as part of the splitup of the old Hacienda Cochela. It has open lands on the altiplano, and some miles away to the north, a separate block of lands allocated from agrarian reform, some adjacent to the mountains rising from the plain. This settlement provides evidence for the strong determinative effects of land tenure on the use of land.

Visiting with Ernesto Chura, ex-agent of PIWA, I spoke to the commune's ex-president Natalio Huayhua Vilca, and to the current president, Teofilo Choque Vilca. They began the *waru waru* in 1990, building five hectares up to the 1992-93 season, between the pampa main site and the allocated lands to the north and west. But the waru remain now only in the allocated lands, as the main settlement is seen as too open. Here too, the men had worked on projects with PIWA, planting over a four year cycle, potatoes for two years, quinua, then barley. Since then, there has been a six year rest. The representatives say that they want to try the raised fields here again, but have obviously suffered from disasters in some years. One major problem is the lack of water, and another is the liability to frost damage. Ironically enough, this is the region in which a major project for irrigation was undertaken, using the outflow from the Umayo lake formed by the River Illpa, but where local farmers have no access to the water as the secondary canals have not been constructed and there are still quarrels over the allotment of water. The raised fields thus depend on rain water, and this is frequently insufficient to protect and nourish the crops. A local feature which has developed here as a result has been the planting of *cañihua* as a crop in the canals rather than on the ridges, as the canals are totally dry for long enough to produce a good crop.

Another major factor, seen often in this region, is a greater attention to livestock farming, which takes most of their time and land. The cultivated land which was visible in Paucarmayu was all alfalfa and barley, for the cattle which are kept here.

Even here, the raised fields have not disappeared completely; one hectare survives, though it is a private parcel of land, not communal. The plot used is one which was developed under PIWA, but after the termination of the project, the owner took the land back and the communal project has broken down. By contrast, there are up to five hectares cultivated (depending on the year and labour availability) on the allocated land under agrarian reform, because this land has remained to a large degree communally owned. A system of use has grown up on this allocated land, which suits the raised field working. There is no permanent settlement there yet, just a barn-like structure, where the men can rest and leave their tools and store crops or seed potatoes. They agree on days when a work party will visit the allocated land, and make a day of it working on the raised fields, with no possibility of being tempted to go home or to work their own patch instead.

Raised Field use here has survived despite a poor record of agency help and a difficult environment. With PIWA, the people had contact with Ernesto Chura first, followed by a series of different technicians who changed in some measure the orientation of the project, and alienated some of the workers.

In this settlement in common with most of the rest there is a sense of projects which have happened and left archaeological relics. One is the set of simple external rural latrines built by a state agency, FONCODES, the Fondo de Compensacion de Desarrollo Social, already falling into disuse. Another is the greenhouse, built by CARE in 1995 and now roofless, with its own water pump. The greenhouse, it is claimed, would be in use but for the capital needed to re-roof it. Money is in short supply. A tractor is working on the next settlement, but it cannot be afforded as its use will cost 25 soles, some £3.50 per hour.

Aricollo

Forty miles south of Puno, on the lakeside near Pilcuyo, a number of communities have developed new raised fields on the land which has emerged from the lake in the 1990s as a result of lake level lowering. At Aricollo, I spoke to Sr. Juan Alberto Challu, from Huayllacta hamlet. His community had worked with PIWA for one season in 1990, building raised fields on what was thought to be communal lands, but was then discovered to be subdivided. In 1992, they approached CARE and built two ha. of fields on land newly emergent from Lake Titicaca, regarded as "common land". This was successful and in 1993 some six ha. were constructed and planted. Some 20 hectares have now been planted on this land and the technology has apparently achieved some diffusion, with individual farmers taking up the technology, and other groups taking up the ideas from the first innovators. In all cases, the lands taken in are along the lake side, where no change to existing crop systems is involved. Here there is a definitely positive view of developments, and CARE has helped move forward from an emphasis on production for subsistence, to some crops such as onions and garlic, aimed at the commercial markets and sales. This example points to a highly successful introduction and diffusion, in the special circumstances of the lake side.

The Bolivian Raised Fields (see Figure 4.4)

Like the Peruvian cases just cited, there are examples in the altiplano to the north of La Paz, Bolivia, of local success as well as of outright failure. In Bolivia, even more than in Peru, it is instructive to note the great difference of the system implied by raised fields, from present day land management. Crossing the altiplano on any of the paved roads, the land is seen to be used largely for animals, for cattle, sheep, pigs, with occasional patches, typically a quarter hectare or less, of cropland but producing crops which themselves are for animals. One good example is the use of mixed barley and oats, a crop that has no commercial value for human consumption, and can always be reckoned to be a crop for livestock feed.

As in Peru, one of the first sites I visited was at Tiwanaku, location of the earliest field experiments in Bolivia, inspired by the findings and writings of the archaeologists. The first of these was the Wilajawira project, led and financed by the University of Chicago and the archaeologist Alan Kolata. The raised fields lie in a shallow valley, a few hundred yards outside the small town of Tiwanaku, and symbolically close to the ruins of ancient Tiwanaku. There is evidence of some six hectares of raised fields, though today, they are entirely covered in grass and used as occasional grazing. I went with a group of students from the University Agricultural College at Tiwanaku, young people interested in perhaps reinstating the raised fields. We talked to an old man, Nicanor Peralta, a farmer working with his son on land nearby. He remembered that the project had been for three years, from 1989 to 1991, and that there had been some success, but that each year had been affected by frost damage. The land was chosen because it was common land, administered by the mayor's office, and the idea of common land and common work fitted well with the archaeologists' concept of restoring an ancient system of farming.

The total failure to use the land again after the project would seem to be a long term judgement on the whole project, but this is not really clear. For three years, the only crop on much of the land was potatoes; in other words, there was insufficient rotation of crops. The location was a frost hollow, according to Nicanor. He also mentioned in our conversation, that this area is one of livestock, not cultivation, and that the only crops really grown locally are barley and oats for the animals, not potatoes. Additionally, the students noticed a number of features, notably that the growing platforms were too wide (five metres) and the channels too narrow (one metre). There was also a lack of drainage and the channels seemed too shallow. There was only a small stream source of water here, and these raised fields could be classified as a phreatic type, i.e. fed predominantly by groundwater. There was also no attempt at reducing frost effects through sheltering. The students agreed with me that from the evidence at other sites in this area, it would be possible to plant the tall *phalaris* bunch grass on the platform edges, which would give some frost and wind protection while also stabilizing the edges. At the College nearby, there are also greenhouses growing *Quishuara* or Buddleia bushes which give good protection from cold winds and grow well here. All these features would have an incisive effect on the success of crop cultivation here.

Ancoraimes

At this little town, 130 kilometres from La Paz up the eastern side of Lake Titicaca, one of the NGOs active in raised field promotion, CESA, had set up a sizeable group of raised fields over the period from 1986 to 1994. A number of settlements round the town had each developed a few hectares in small projects; at Llojllata, four hectares; at Ispaya Este, six ha. At Lacaya, eight ha.; at Lojro and Cachi, eight ha.; and at Ancoraimes itself, four ha. The history of the project is like that elsewhere; development while payments and the project were active, then a halt. Projects lasted a year or two extra here because the national raised field agency, PROSUKO, took over the work of CESA in 1993 and carried on with it for a while. Today, there are between two and three hectares still in use.

I looked in detail at the Ancoraimes sites, a mile inland from the town which sits near the lake, and in a well watered valley, on a wetland site which needed work to make it usable for crops. A special crop was selected here, the onion, for which there was a good market in La Paz. Even today, small patches of onion are cultivated here. The land itself, as at Tiwanaku, was communally owned and normally rented out to individuals as needed. Again at Ancoraimes, the construction seemed to be irregular, with canals only a metre wide, apparently simply to drain the land and forgetting the effects of micro-climate, and the platforms too low at some 50 centimetres. The farmers who took me to the site, though they had not worked during the project, knew that much help had been given, in tools, seed, technical assistance, and money for work in the beginning. As elsewhere, they would be happy to begin again, given a new project.

After this visit, I was able to read a report written in the first years of the raised fields projects (Intercooperation 1992) which focussed on Ancoraimes as it had been one of the first areas for any attempt at reintroduction of the technology. Already at that time, the experts appointed to look at field projects noted irregular construction; high levels of iron oxides washed down into the project soils; and construction of the platforms to dimensions which were too narrow. But they also noted variations on the theme of raised fields, suggesting that no single, simple type of structure was relevant to all areas. On poorly drained land, they found the use of *huisu kollu* was important. These structures involved raised fields, and on top of them, raised platforms at right angles, to improve drainage still further. Another technique was the use of *tara suka,* again in wetlands, a technique for cultivation and drainage of wetlands, in which one to two metre platforms are constructed with a slope towards a central ditch, imitating to some degree the natural tree structure of river drainage systems. The central ditch is the main drain like a river, collecting and extracting surplus water.

Overall, Ancoraimes was a special case with different conditions and crop potential. It was abandoned by the agencies no doubt because of its relative inaccessibility for managing projects, but its physical potential was as good as anywhere else. As elsewhere however, it was abandoned by the locals as soon as the project team left.

Achuta Grande

Yet another different situation obtains in the main areas of action by PROSUKO. On one day, I was taken (on the back of a 175 cc. Japanese motorcycle belonging to the PROSUKO agent, Pedro Marin) to talk with a few of the farmers linked to the agency, at Achuta Grande, some five kilometres west of Tiwanaku. A success story is that of Florentino Quispe Choque; this farmer is relatively wealthy and entrepreneurial. He is an ex-hacienda worker, who was given ten hectares of land when the great estate of Achuta Grande some 1200 hectares, was broken up in the 1960s, following agrarian reform in Bolivia; of his ten hectares of land (25 acres) half a hectare is in *suka kollus*, the raised fields. He has a carefully managed rotation of potatoes, broad beans, quinua, and barley/oats, followed by a rest period. He had originally contacted PROSUKO as an individual, simply from observing neighbours and noting their projects with the agency. He has worked with them over three years as an individual, depending mostly on family labour which means five men workers, plus some help from hired labour for the initial construction, and for canal cleaning which he does do. This is a relatively large farm for the post-reform period, and it may be noted that Florentino does not depend on the success of the raised fields. He has 40 sheep, and other livestock, and most of his land is cultivated, yielding good crops of fodder for the animals.

Florentino is one of some 30 farmers at Achuta Grande, in similar circumstances, who have all had projects with PROSUKO, out of a total of some 120 families. For each of them, the *suka kollus* is an additional form of land management, something which gives them diversity in cultivation, and for most of them, the land is cultivated, rather than left to grass. Unusually, this piece of the altiplano is cultivated rather than left to livestock extensive grazing, perhaps because they had experience under the previous hacienda of producing field crops. The lands used for the raised fields are generally lands of poor drainage, if not downright swampy ground, and the raised fields are definitely seen as a solution for bringing these patches of underused land into cultivation.

Limitation of the Use of Raised Fields

One interpretation of what is happening, is that rather than abandonment, or lack of interest generally, as would be a first impression, there is a general limitation of the use of the fields to specific physical environments and regimes over time. In many places within the community of Caritamayu, at the Paucarcolla site, and in Bolivia, at Tiwanaku, at Ancoraimes, and at Achuta Grande, the choice of site is a wetland or swamp with little or no current use, and the use of the project has not been to introduce a technology which would be adopted over a broad area, but an adaptation to a special situation where drainage was urgently needed. In the case of Aricollo, the special situation was the lowering of lake level and the uncovering of new land, allowing a communal effort to be made before the land was claimed by private owners. Is this a failure or a success? Obviously, the answer varies from the point of view of different rural communities, and it may not be regarded as a great success from the viewpoint of agencies intent on general introduction and adoption.

The other obvious feature is the apparent abandonment of the raised fields in most areas, though whether this is truly abandonment, or merely a long fallow, cannot be determined for many years. Making a generous allowance and regarding all the unused fields as long fallow, it is still the case that the adoption of the raised fields is on a smaller scale than hoped for. But perhaps the agencies were themselves wrong, and what can be observed is in fact a success, despite all the criticism.

At the beginning of this chapter, the serious constraints imposed by demography in the region were mentioned and the heavy pressure of population is central to the actions or inaction of local communities. When families are so close to failure, their unwillingness to commit strongly to any new technology must be intelligible. A correct response to innovation must be caution, as any mistake will leave them worse than before the project. Another feature, applicable in some of the communities visited, is the relative unimportance of arable farming in their system. Passing towards Lake Titicaca from La Paz and only a few miles outside the city, the land is devoted to cattle, not to crops, for mile after mile. Those crops grown are fodder barley and oats, often mixed, and not food crops. Along the lake side in the Peruvian sector, there is a narrow strip of largely arable land, but even this peters out a few miles inland, or even within one mile where high land and steep slopes approach the lake. On this rougher land, animal husbandry again dominates over arable cropping. The importance of the animals has been given little recognition in the programmes for raised field introduction.

6 Agroecology and the Raised Fields

Introduction

If raised fields technology is seen to have limited results in terms of economic returns, the maintenance of social structure, or even as a risk-reducing technology, then it is still possible that the programmes for re-introduction have a justification, because they support sustainable land use, agricultural systems which can be maintained over many generations, without damage to the existing stock of human and natural resources for agriculture in the Andes. If raised fields are compatible with agroecological principles, we could justify the programmes as part of a sustainable development package for the region.

Agroecology is the applied science which links from agriculture to sustainability in farming and the maintenance of ecosystems. It is concerned with soil erosion and the depletion of soil nutrients, with the conservation of wildlife, with the management of water resources and the control of pests and diseases by working with natural forces. For some writers, at least in discussions of agroecology for Latin America, it includes the social processes and practices which affect and often control the purely physical and technical elements of a farming system.

The question which is raised in this chapter is whether in fact raised fields and associated technologies do help sustainability by using agroecological principles. At first glance, the answer might seem to be an automatic yes, since the technology involved is known to have been widespread in ancient times, and to have been employed over a considerable period of time, without use of modern farming inputs such as heavy machinery, fertilizers, chemical herbicides, and pesticides, which rely on outside sources and whose long term use is likely to be damaging to the environment. But because conditions have changed, it cannot be relied on that current re-introductions will faithfully copy the past, or that they will fit in with a present day system that is in other respects ecologically sound.

Before we go into the details of how agroecology applies in the Andes, it is worth pointing out again that many of the changes occurring in Bolivian and Peruvian agriculture are hostile to the maintenance of traditional farming, farming which itself was often diversified, stable and sustainable, and which in the best examples contained many positive agroecological elements. Traditional farming, given the holistic nature of traditional society, was dependent on a balanced linking of social, economic and technical elements, damage or loss of any one of these would cause a general decline. The break up of the ancient communities, the *ayllus*, is one such major kind of damage over recent decades. Ancient communities are known to have combined mountain land, slopes, plains with wetlands, and valleys at lower levels, utilizing the resources of all these. These communities have been broken up by administrative subdivisions from the post-Independence period, and then the remaining communities totally atomized by agrarian reform and counter-reform legislation prioritizing family farms and individually owned land. Finally, today the norm is for each family to own a single compact area in one kind of environment, and this itself may mean over-exploitation of resources through lack of alternatives.

Apart from these specifically agricultural land changes, modernization of the society and economy of the Central Andes puts increasing pressure on farming. Part of the change is the break up of the ancient communities themselves, which links to the land changes just mentioned. If the community breaks up, then it certainly cannot jointly manage the complex farming systems of days gone by. But more subtle changes are also observable; genetic erosion, the gradual loss of old cultivated varieties, is observable throughout the Andes, but most notably in areas where modern, market varieties have been encouraged and introduced. Production for the markets, rather than production for subsistence, also moves generally towards modern ways of farming which do not allow the use of time and space in the same ways as before. Crops must now be produced for markets when there is a market demand, not when they can most naturally be produced. And crops must be produced near to the cities where they are to be consumed, not in remote villages. These kinds of forces act against agroecological principles.

Agroecological Principles

For a summary view of main aims in an ecological farming system, we may take those summarized by Altieri (Ch. 3, in Uphoff 2002).

1) one of these is the minimization of energy losses in the micro-environment (micro-climate, soil environment) of plants. This, from what we have already seen in the description of the technology, is certainly well adhered to by the raised fields;
2) the enhancement of biomass recycling, so as to make nutrients available to plants when needed and avoid losses;
3) the provision of favourable soil conditions for plant growth, by managing organic matter and enhancing soil biotic activity;
4) the diversification of species and genetic resources in the agroecosystem, over space and time;
5) the enhancement of beneficial biological interactions and synergies among the components of agrobiodiversity, so as to promote key ecological processes.

Obviously, these principles have different applications in different areas of the world, but the overall idea is the coordination of the different elements of the farming system to maximum effect, in many respects aiming to imitate the structures and functions of natural ecosystems.

Energy Retention

Taking the first principle, raised fields technology provides a definite element of minimization of energy losses through its modifications of micro-climate. Construction of the fields has been shown to have an important effect in limiting the damage due to frosts and freezing air, which is a constant worry for local farmers. This may be the most important element in the raised fields technology. The warming effects of solar radiation may also be enhanced by the fields, since they present a greater area for the sun to heat than the flat plains. This allows the soil to heat up better and help root and plant growth. Apart from micro-climate, water is also better conserved by the raised fields, as it is managed by controlling access to the ditches and flow through the raised field systems. However, this is a minor consideration and raised fields are not managed like irrigation systems where water flow is the key. Soil management also helps in raised fields, at least in theory, since the idea collected by archaeologists was that the soil from the ditches could be recuperated after several years, as fine mud enriched by organic wastes from the ridges; but this idea has remained at theory level, though its potential may yet be realized.

Biomass Recycling

Our second agroecological principle, biomass recycling, cannot be observed in the field projects reviewed in Peru and Bolivia. Digging out the ditches, as just mentioned, would have been the best way of accomplishing this objective, bringing back both topsoil and organic matter to the cultivated area, but it has not been attempted. According to villagers asked about this, there is awareness of the potential, but the amount of good mud accumulated over ten years has been insufficient to justify the efforts of digging out. In a few cases in Peru, some use of the ditches themselves, in the dry season, for cultivating *cañihua*, has been the only way in which the organic matter has been recovered.

In temperate latitudes, a major recycling process on farms is through the use of animal manure. But this is not found on the altiplano in general. Most communities keep animals, but on separate land, not on the arable land. Some dung is left by animals grazing over stubble land which is land left after crops, for several years, but this is circulated through the weed growth on these resting fields, rather than through the direct circuit of manure onto ploughed land for next season. In some areas, such as that around Huatta, shepherds with no land are found and these people are in conflict with local farmers because they may let their flocks run over any attempt at crop raising on the flat land. Their claim is that they always had free run over the plains, since this land was not used for crops in olden times.

Even more depressing is the use of dung for fuel, seen north of Juliaca and near Acora, on farms with no access to nearby woodland and no money to buy kerosene for burning. The dung is collected from the open land, dried on the adobe walls around the kitchen gardens, and used when dry for cooking and a little heat in the house. Such use repeats the uses made in the poorest parts of India and Africa and constitutes part of the problem of soil impoverishment and erosion, leading to poorer crops from year to year.

There is a system of crop rotation, which does manage to retain some biomass through the cycle, notably on the areas managed in *aynokas*, (an aymara word, in Quechua replaced by *manta* or *laymi*), effectively three field systems comparable to those of medieval Europe. On these, there is a rotation of crops for any one great field (in which each farmer has one or more strips for his use), typically potatoes, followed by barley, quinoa, and legumes, usually broad beans, and then a fallow of four to eight years. Such rotations are found throughout the lakeside lands, at elevations of 3800 metres (the lake level) to the upper limits of cultivation at 4300 metres. The fallow helps in conserving biomass and soil protection. The advantages cited by Hastorf in a work relating to the Peruvian Andes generally, and especially the Mantaro Valley (1993, p. 30) are:

1) that animals can be controlled without walls and fences;
2) that animal dung is used as fertilizer;
3) that the fallow allows soil fertility to be recuperated over time and;
4) that the *aynokas* allow irrigation systems to be properly located.

In the area around the lake, the *aynokas* are a system of the altiplano, not of irrigated land. Indeed, *aynoka* systems here contrast with the continuous use of lower lands, at 3000-3800 metres, where irrigated land may be under continuous use (Schulte 1996, pp. 51-70). However, the *aynokas* have never been integrated with the use of raised fields, but treated always as a distinctive type of system. We should also note again, that the long resting period is not what was expected or specified by the raised field promotors. It is true that the fallow period of rest is being shortened in some areas through pressure of population, but the principle remains and from discussion with farmers throughout the altiplano, a long fallow is what is taken as the norm.

Soil Enhancement

On the third principle of agroecology, enhancement of soil conditions for maximizing plant growth, particularly with reference to soil biota, the raised fields have only a limited contribution. Addition of organic matter, it has already been noted, through manuring, is only done incidentally or accidentally. Use of green crops which might be ploughed in, or used as a mulch, is similarly virtually absent in local traditions. In the last five years, the agencies themselves have introduced the use of alfalfa as a high value crop which may be used for feeding animals, and which will also enhance the nitrogen levels in the soil. CARE and PROSUKO, as mentioned previously, have both advanced free supplies of alfalfa in these latter years, as they have realized that the raised fields have gone out of production after the projects some to an end, and that the farmers when queried about this cite the lack of fertility in the soil as a reason for abandonment of the land.

There is no tradition in the region of the use of compost or any vegetable remains to enhance soil conditions. It is likely that there is some improvement to soil aeration through the construction of the raised fields. Especially on areas near the lake with poor drainage, this feature is of importance, allowing better root development and heating of the soil, but it is unquantified, and the most important feature remains the overall loss of soil fertility.

Diversification

The raised fields add nothing in terms of species diversification and conservation of genetic resources. The agencies promoting the raised fields have not generally attempted to add to diversity, though in the case of CARE, they have been involved in the promotion of quinua and cañihua crops which had become kitchen garden crops rather than field crops in some areas. CARE and other agencies have also fostered the cultivation and adoption of *maca*, a root crop like a small turnip, which has been used as a health food and sold abroad. But these elements are minor parts of the programme and have scarcely changed the overall situation.

This situation, especially with respect to genetic resources, is disturbing. There is evidence of a declining level of variety in the main crops grown, what is now commonly termed a genetic erosion. For example, in the very traditional farming area of Charazani, northeast of the lake, where Michael Schulte's group of researchers worked, he notes (1996, Ch. 9) that individual communities have around 50 to 60 different varieties of potato, while individual families may also have near fifty. On the altiplano to the south according to Schulte, more affected by modernization, the number is reduced to two to three varieties. In the region around Puno, the chief project officer of CARE, Alipio Canahua, with other researchers, in the interval 1991-2001, gleaned detailed information about the genetic diversity for the main field crops, which has shown that in the high lands above 4000 metres, the *Suni Alta*, 132 different cultivars or varieties of potato are used, and 39 of quinua, while around the lake in the area closer to modern urban influences, the corresponding numbers are 63 and 35. This information was reported in a paper delivered to the 9th Biennial Meeting of SEPIA at Puno on 24 October 2001. The influence of closeness to cities and markets is undoubtedly the cause of this genetic loss. Although Canahua does not explain the origins of the diversity, they must lie in the remoteness of each village and its isolation from its neighbours, without much interchange through sale or barter of crops. This isolation allows a genetic drift over centuries, so that the successful local varieties match the local ecological niches exactly, and this means that each village's varieties of crop differ from the neighbour communities, and even each farmer's crops will differ from their neighbour. All this is lost when modern, high-yielding varieties are introduced from outside and promoted by outsiders.

This kind of criticism about genetic erosion is not solely the problem for raised fields. Throughout the world, programmes of modernization cause similar losses of diversity. The cases recorded by Ruddell (2002) are relevant as they describe programmes which are thought of as ecologically desirable, in the Andes of Bolivia and Peru. These programmes were introduced by World Neighbours, a USA-based NGO. In Bolivia, a highly evolved technique was used for identifying good varieties of potato in northern Potosi Department, by asking the farmers themselves to conduct trials in small areas of their farms, rather than the more traditional method of using agricultural experiment stations to establish best practice under field laboratory conditions. This did ensure local acceptance and the use of varieties which were most productive at each site. But it did lead to two or three varieties becoming the most popular and most advertised among the farmers. As Ruddell comments, in the first years of the programme, Sanimilla was the favoured variety, two years later Waycha and Qorosongo, then Colombiana. Adoption of these varieties means leaving aside and eventually loss of some other varieties whose genetic qualities could still be important. In Peru, Ruddell comments on a variety of comparable programmes. One was for the elimination of barley rust. A high producing, rust-resistant variety developed by the National Agrarian University at La Molina, was identified and brought in, and proved to be a great success. Local farmers themselves were used as promoters and propagandists for the variety, which was essential because the grain could be stored and would stave off famine in bad years. What Ruddell fails to note is the number of local varieties which may have been lost, and which might be relevant to the many varied ecological niches in the Andean environment.

We shall return to the subject of diversification in the next chapter, but from a wider, more inclusive context, because diversification is a realizable aim for rural improvement programmes.

Biological Interaction

The final principle advocated by Altieri is that of promoting biological interactions and synergies among elements of the agricultural system. From what has already been said this principle cannot be well sustained by the raised fields as now practised. Because they exist as an isolated system, on land set especially aside for the purpose, the interactions with the remainder of the farming system are slight. On the *ayllu*, the complementarity between the different fields (i.e. spatial integration) is supplemented by temporal integration, the rotation of crops over time. The raised fields do have a rotation, but are conducted at single points in space. Within the community space, there was also integration between the use of the high alpine grazings, the terraced slope lands, the open altiplano grazings, and the cultivated fields, but this was only possible because the land tenure, especially communal ownership and carefully defined rights for use of each element, were specified in advance. The raised fields did not enter into this system.

There was always room for a potential realization of synergies between different elements in the raised fields, if their use could be made supplementary to that of the other elements – open plains grazing, terrace farming, fishing etc., but by the aims and scope of the projects for raised fields, this could never be organized.

Reviewing the five principles enunciated by Altieri, the commentary so far has not been very supportive of the concept of raised fields as ecologically based. This is worth noting because Altieri himself, in a previous text (Altieri 1995, Ch. 7, pp. 145-78) had used the *waru waru* of Peru as an example of an ecologically based agricultural development programme. Such programmes, he thought, should have the aims of (1) increasing productivity of land and labour; (2) using conservationist methods of soil and vegetation management, and rationalize the use of chemical inputs; and (3) work within an appropriate national background of prices for farm goods, taxes on farmers, land tenure and land access, and use good technical assistance. It is of course true that modern high technology farming also shares these aims in many cases, so the differences boil down to details on the conservationist management of soil and vegetation, avoidance of inputs from modern science, and the like. What Altieri has in mind through this discussion, is the contrast with Green Revolution farming, using the best farmers, the most advanced modern technology, large amounts of capital, maximum production of a few crops.

The raised fields certainly do not represent a Green Revolution approach, but they are shown in this chapter to fall short of an agroecological one. The projects on *waru waru* that Altieri used for his supportive comment, were those of CIED, the Centro Internacional de Investigacion para el Desarrollo, in Puno, dating from the late 1980s and up to 1992. These were not followed on by CIED, as like most other agencies, they moved on to other programmes after a few years. From conversation with Jorge Reinoso, director of CIED, in July 2000, it is apparent that this agency could not go on supporting what had set out as a totally agroecological approach, involving only traditional farming tools, no tractors for earth moving, and no chemical inputs in fertilisers or herbicides. While CIED projects had shown the potential for reintroduction of raised fields using traditional technology, this agency took the view that in fact a combination of modern inputs with raised field principles had the best chance of increasing farm production, and in any case preferred to leave such an approach to other agents. Long term viability of an ecological approach was not really tested.

Andean Versions of Agroecology

We may come back to the definition of agroecology used in Latin American writing on the subject, a definition which includes the actual social practices of the people involved, and the ways in which these may be integrated into the farming system. The case may be made here that there has been little evidence of integration of any existing social customs or traditions with the re-introduction of the raised fields. Some of this failure is to do with the very nature of this farming system; thus for example, the integration of the aynoka system, involving common decisions on the cultivation and management of large areas of the village land in the manner of the European medieval three field system, would be difficult to make with raised fields. No raised field area is large enough to be treated as a complete field with a single crop, and the raised fields, when in use, are relatively high intensity requiring much more effort than the great fields of the aynoka. Communal work systems have been adopted by all the agencies, when the fields were first built. The ancient *minka* system of bringing together a number of farmers in order to complete a job like harvesting or sowing of a field, was used widely to complete the construction of the raised fields. This involved a payment in food and drink for all those brought together, or in products which the farmer may have available such as potatoes, onions, or sheepskins. *Minka* could also involve payments in money, and this was included in the early attempts at introduction of raised fields. Later the money aspect was dropped by the remaining agencies, as an unsatisfactory method because the work was done by farmers or labourers who became dependent on projects for their income, and the aim of re-use of an ancient technology was lost sight of.

But some kinds of ancient reciprocity were never accessible to the projects, because they operate at a different level. The most basic of these, the *ayni*, involves a simple exchange of work or some social activity, between community members, and the balance is kept between these members by a reciprocal offer of activity in the opposite direction. Besides farm work the *ayni* may include social visits and feasts to celebrate events such as a marriage, a birthday, or a death. It depends on continuous presence of the actors, so that migration out to the towns, and even seasonal migration to the coast, tend to break up the system. For something like raised fields, the unlikelihood of anyone being able to reciprocate in the same coin means the system does not operate.

What cannot be retained are a host of social practices which are of ceremonial nature, serving to bond the members together and add symbolism to any economic act. These are practices which have no direct economic meaning but give strength and solidarity to the society which manages the economy. Andean farming, as with other traditional peoples, exemplifies what Pearson (in Polanyi 1957, discussed in Morris, 1981), called an economy operating within a social integument, so that society controls how the economy operates, and social rules must be respected. We may cite from detailed studies of the social organization of farming in the Bolivian highlands (Villaroel, Salazar, and Delgado 2001, Rojas 2001) who show the complexity of the cultivation of potatoes. Rojas's account relates the ritual "raising" (*crianza*) of the potato, at Chorojo, on the edge of the altiplano. A first statement is that the cultivation of potatoes is something which involves attitude, and that it must be done in an atmosphere of respect and goodwill. The first decision is a common agreement, in the Spring, in October or November, to go out to the fallow field, each family making its prayer and offering to the Earthmother, the Pachamama; this is to ask her permission to plough, which is then done with a pair of oxen. Further prayer is made to allow sowing, when an offering is made in the field invoking the Pachamama and the ancestors and spirits of the crop itself, in order to achieve a good harvest. At the time of Carnival, from mid-February to mid-March, there is a potato festival, when another offering is made, and when *chicha* (maize beer) and other drinks are taken, in a *challa*, an offering to the Earthmother. This is followed by a dance with music of the charango, a small guitar, or accordion, and this is attended by the *alcalde de campo*, literally the mayor of the field, who organizes drinks for all and then a competition in *taquipayanakus*, songs and tales of the locality, between different families. In March or April, the harvest is begun with another prayer and offering to *pachamama*, and the first harvested potatoes are blessed by the grandparents. Some families play games at the harvest before taking the harvest home.

The study by Villaroel *et al.* (2001) comments on these rituals, and adds others from the same zone. One interesting one, which was also observed by this writer, is the making of meteorological observations, in the case cited by Villaroel, in early August, a significant timing because the observations are centred around the prediction of weather for the coming growing season. Peasant traditions use the observation of some animals, birds, and weed plants, to give a hint as to the nature of the coming season. Near Tiwanaku in 2001, Jacinto Quispe Choque told me of the study of clouds in August. Cloud in the early days of August meant early sowing, clear skies meant late sowing because of the frost danger. The mating of foxes before St. Michael's feast, in September, meant early crops. The early or late dates of flowering of the weed plant, *huaycha*, gave a positive or negative indication of the crops later. He also gained information on the weather to come from talking to the llama and alpaca breeders who live in the mountains west of Tiwanaku, at over 4000 metres. These people come down to the town of Tiwanaku for the Sunday markets, to exchange their wool, fleeces, and salt from saltpans, for grains and potatoes, and at the same time exchange information based on their own observations of the night sky and of their animals' behaviour.

Such elements as these lie outside the modern conception of agroecology, yet they are part of a wider traditional knowledge which seeks to maximize the efficiency of use of resources. Much of this knowledge is in the process of dying out in the countryside, irrespective of the technology being applied, but it might be thought that any attempt to resuscitate an ancient local technology would make major efforts to associate with traditional social elements. As noted in Chapter five, the early Wiljahuira programme did have an attempt to associate traditional social practices with the raised fields, but this effort, over a few years and inspired by foreigners, came to nothing.

Some Afterthoughts

The comparison of some aspects of the raised fields programmes with principles from agroecology is not conclusive, as some elements are seen to be positive and others clearly negative. However, the overall direction of the programmes is definitely not as strongly ecological as might be hoped or expected. There are partial, small links to traditional farming and beneficial practices, but some of these may actually be countermanded by a modernization effect that any programme will bring in. Dependency on outside agencies of any kind must tend to break down a system which relies on multiple and complex inter-relationships between social, technical and economic elements.

It was pointed out earlier in the text, and in the beginning of this chapter, that the overall context for retention of any traditional farming system is weak, since changes towards atomization of the communities and replacement of communal land and communal working by private small farms has been happening over recent decades. In theory, the raised fields could be constructed and worked by individual farmers, even in quite small plots, as the system does not depend on massive infrastructural works like an irrigation system with all its main and branch canals. At Huatta, several of the farmers who I questioned about this were of the opinion that even though there was no communal interest, they could restore the fields themselves. Psychologically however, this would seem to be a huge challenge, requiring much work from the individual or the single family, and without any guarantee of the results meriting the investment of time and effort. All the projects to date have in fact depended at least to some extent on group activity, on some kind of organization and a number of families working together.

There is one positive aspect to be drawn on here. It appears that there is an agroecologically desirable system of use for the raised fields, already understood and used by the farmers, but not recognized or accepted by the agencies, or at least, not in accordance with their aspirations. This system is similar to the one which is already practised over the terrace lands around Lake Titicaca. It consists of cultivation for a period between one and four years, and then a long fallow of up to ten years or more. This long fallow is ecologically sound, as it allows weed vegetation to build up and protect the soil. It also allows pests which infest the crop, particularly the potato crop, to be eliminated. These include the *gorgojo de los Andes* (an infestation by large worms), some nematode worms, and other pests.

After a number of years, a new crop, when it comes, will find nutritious soil and a low level of the pathogens which would reduce yield in the first years of a new round of cultivation. Such a long term, low intensity system is certainly not what the agencies have anticipated, and not what was envisioned by the archaeologists who gave the first inspiration for the re-introduction of the technology. High yields and a constant intensive use of the raised fields were the basis of their calculations, and of the costings which showed a strong economic viability for rebuilding the raised fields. Long fallow periods would make the economic and production basis for the programmes less obvious. But perhaps this is the real future for the raised fields, and what is acceptable to modern small farmers.

7 Some Final Lessons: Diversification and Integration of Farming

Introduction

On the face of it, there has been a double failure in the efforts to reintroduce raised fields technology to the Central Andes. This failure consists, first, at an intellectual level, in the lack of conclusive demonstrations of the value of the raised fields in standard economic, social and environmental terms, as we understand these in the west. A second failure is in the programmes for introduction. There has been a lack of permanent use of the fields in intensive farming, for virtually all the cases studied both in Bolivia and Peru. At the end of each project the land has been left, whether abandoned or put into a long term fallow.

My thesis here is that despite these failings, the idea and the overall programme for introduction has brought benefits and will do so in the future and the criteria of failure are themselves suspect. Here, we review the reasons usually put forward to explain failure, which seem to account only in part for the poor levels of apparent adoption. Then the study goes on to consider the ways in which diversification operates in the Andes and what this may mean for any new technology. The raised fields can operate as one element in this diversification and as a means of diminishing the problems of an uncertain environment. In relation to the archaeologists' dream of reinventing the past, of bringing back a long lost technology which was the secret of supporting a strong, large population, there has perhaps been a failure. But for the farmers around Lake Titicaca, the raised field programmes may be presented as not a failure at all, but an important addition to their technological and physical choices.

Mention was made in Chapter four of the collapse of most rural programmes after a period of initial euphoria, and the similar fate that seems to have happened to the raised fields, so that it might appear that the only result of many years of effort has been the research results which show the strong physical and technical viability of the idea. It must be noted that the publications and internal documents of the agencies have always shown a positive benefit from their efforts, but the case has been made that their measures have been inadequate for the task. Standard measures of success include counting the number of communities to have areas put to raised fields, the number of hectares actually being cultivated in raised fields, the production and productivity of the fields which are in use. For most of these measures the only data released by the agencies are those relating to ongoing projects at the time of measurement, and there are no long term measures. Short term, in-project measures mostly come out with good results, sometimes spectacularly good, but given the widespread "abandonment" of the fields, or what may be regarded as long resting times, the programmes must be evaluated, in standard western terms, as expensive in terms of costs per hectare of land and per year of time since first initiation. To take the number of communities actually involved currently in the programmes, these are very small in Bolivia, with only about 16, and in Peru reach about 100. There have been hundreds of projects and communities involved, but many retain only the memory and the forms of the raised fields, and their use in the future is unknown.

But to judge the whole experiment as failure is to misinterpret the thinking of people in the rural communities. There have been high costs involved, as both in Peru and Bolivia major efforts have been made using NGO and government money to encourage the use of the fields. The costs however include those of initial establishment of the raised fields, not their permanent maintenance which might be less onerous over time. The case may also be made that very high initial costs are inevitable because of the need to convince a rural population of the value of a whole technology, which will be higher than those for an individual crop. On the other hand, the benefits are hard to estimate, because they permeate through over a number of years, and may only be realized if the period of study can include major disaster years, such as those of heavy frosts or flooding, when the raised fields were in use. Even in the absence of disasters, the measure should take place from one period of several years use to the next. This might mean studying benefits over a 20-30 year cycle.

Intellectual and Practical Failure

At the intellectual level, research done by the agencies for reintroduction of the raised fields revealed what were the problems for any programme. None of the economics arguments tackled the long term, seven or ten year cycle of land use. None of the social studies tackled the essence of Quechua and Aymara culture and the need for the economy to be viewed as within this society. Ecologically, though there were no conclusive studies, it could not be demonstrated that raised fields helped manage the lands of the altiplano. Only the firm evidence of the archaeologists could be adduced to show that this had, at one time, been a successful technology.

At the practical level, the projects had been successful in finding how to rebuild raised fields, how to get communities started and to carry out a four year project for reconstruction and use, but as mentioned at various points in this book, nobody could demonstrate a continuous use of raised fields in specific sites over a period of seven or ten years.

In Chapter four we looked at some of the factors to be taken into account in explaining what has happened. These included the physical site and setting, the economics of construction and production, the land tenure situation and social context of the programmes. None of these gave a final or complete explanation, though at least the physical and biological grounds for the whole experiment were seen to be very positive. Good crops can be grown with this technology. In other kinds of projects of farming aid round the world, the choice of site for the projects has been found wanting, but here the sites are usually good, in fact most raised fields are in the very places where traces of ancient raised fields have been found. This has even been a way for agents to gain farmer confidence, by saying that the new raised fields must be on the right site because traces are still there to mark the previous efforts.

The fields also seem to have been constructed with understanding of their use in combating flood or drought conditions, again because the ancient patterns can be seen and followed. The all-important labour effort has been enthusiastic and considerable, and there has been, at least in all later projects, considerable local participation in the decision making and building of the fields. Land tenure has been a major source of conflict through the recent decades and some failures or refusals to accept raised fields have been based on the uncertainty about security of tenure for the farmer. But this cannot be a general factor everywhere and the increasing emphasis on dealing with families rather than whole villages avoided some of the problems associated with break up of ancient communities.

We cannot know the exact conditions under which the raised fields were developed, and so one line of analysis of the apparent failure may be to argue that the physical or social environment was different when they were built. Some writers who have examined this matter, Kolata for Bolivia, Kervyn for Peru, assert that the whole system of Andean technology operated in conditions of:

1) a dense rural population, providing a large labour force, and with Boserup style demographic pressures under which innovation stems from the need to survive;
2) use of the produce from the fields very largely to feed the local population;
3) a central organization like a state, with considerable powers of coercion, to ensure proper hydrological control of the major resource, and management of labour for the massive works needed.

But these conditions, whether or not they existed in pre-Columbian times, are not perhaps necessary for the success of this technology. Erickson makes an excellent case for the very early evolution of raised field farming by lakeside or wetland farmers, when there was only a tiny population in the region and no demographic pressure. He also shows that the diversity of the raised fields in form and even function, and the irregular overall pattern, betray not a central state control, but many small communities, acting alone and independently, for much of the time in which the fields were built.

There may have been some difference in the physical environment, and the evidence is that major changes in climate have occurred across the centuries, but there is no real indication that today's climate is unfavourable for the technology. Certainly, the fields have proved their worth in disaster years when there is abnormal flooding or cold, and there are many of these on the Altiplano.

More importantly perhaps, is the lack of understanding for the exact mechanisms by which fertility was maintained in the raised fields, and this knowledge must be a key to future success. Most farmers do note that the yields decline with time, and that the land needs resting; an important element of cultivation in many areas of the Central Andes is the use of a long fallow, up to 10 or even 20 years, which is easily interpreted as abandonment by western observers, but must be understood as an integral part of the system which seeks to use low fertility soils for cultivating crops which need large amounts of nutrients (Pestalozzi 2000). In ancient times it is possible, that apart from long fallow periods, high fertility was maintained by constant additions of animal and vegetable manure which are not now made available.

Another whole set of factors used to evaluate are the economic ones, but again (Chapter four) we have noted that the evaluations in economic terms are very suspect. In particular, the measurements have been restricted to the project period, which at a maximum is only four years, when in reality, a period of 20-30 years could be the optimum. In addition, the measurement has assumed positive labour costs in all cases, which is not necessarily true. Reference to the work of Chayanov has been made in order to show that in areas of purely farming population and crop production, there are long seasons with little work available, and that these could be occupied with construction of raised fields. The economic case is unclear and incomplete.

In previous chapters, some discussion has been made of the Andean approach to farming, based in multiple technologies and products, to minimize risk and ensure survival. Any innovation should be able to be integrated with the rest of the economy, and indeed with the local society. We have also noted the need in any long term approach, to use ecological principles and make farming sustainable. We may now return to these ideas.

Diversity and Diversification

Many writers have focussed on the diversity of jobs, of resources, of products, that are used by small farmers in marginal areas around the world. In the case of the Central Andes, local writers use the term "Andean Technology", giving the idea of a whole set of special elements which are linked together and are unique to the region.

There is no need here to investigate all the different ways in which this andean technology is expressed, but along with the raised fields, there are a number of other engineering or land moving projects which may be mentioned and which sat alongside the raised fields in the past. To take the most evident and relevant example, terraces on steep hill slopes cover huge areas in the Central Andes. Terraces themselves are almost a lost technology, like the raised fields. About half of the Peruvian terraces are apparently abandoned, some 750,000 hectares by one estimate (Burga 1986). In some areas, like the Colca valley, over 60 per cent are abandoned, and of the higher terraces over 90 per cent (Denevan 1986). CEPAL suggests a total area of terraces of the order of about two million hectares in Peru alone (CEPAL 1989). There is no way of knowing directly their integration into the rural economy with other elements such as raised fields, but they occupy higher ground with often steep slopes, and they certainly date from considerable antiquity (Denevan 2001), so that they may be assumed to be a part of the productive system alongside the raised fields. This is the assumption of Donkin in his review of these features throughout the Americas.

Terraces were themselves diversified, and are classified in detail according to form, function, use of water etc. Some were dependent on rainwater, others were part of an irrigation system, all had arrangements for dealing with drainage water. Because of their location and soil types (usually they were warmer in cold spells than the plain, because of air drainage, and they avoided waterlogging and flooding at all times) they complemented the fields on the flat plains. The fact that today, slope land in terraces is used for crops such as potatoes or barley in the altiplano lands, and the flat plains for natural grazing, is indicative of the differences and the complementarity possible. Terraces must also have been used intensively, for otherwise it is difficult to see how the huge amounts of work on their construction could be justified. A considerable effort is needed to keep up terrace walls, to maintain surface slopes, to prevent gullying, to ensure drainage of excess water.

Like the raised fields, the physical advantages conferred by terraces are complex and important. They retain the soil itself on steep slopes, and allow its accumulation in deep fertile layers. They conserve water and make it accessible through the deep soil, to plants, checking its downward movement and guiding it horizontally along the terrace rather than directly downwards. Some terraces have channels to lead down from one level to another, in other cases the water seeps through walls and soil alone. There are thermal effects, through the strong heating of terrace surface and walls in day time so that some of the heat is conserved overnight. Increased turbulence of the airflow, due to the roughness of the terraced hill slope, means that downslope movements of cold air, a blanket of which develops close to the ground through outward radiation at high altitude, are disturbed and broken up.

Today, the massive abandonment of terraces is leading to the breakdown of the system, and particularly to the increase in soil erosion which will eventually destroy the terraces themselves. There is good proof of the erosion control function of the terraces in Peru (Inbar and Llerena 2000), showing that good terraces almost totally eliminate erosion, while abandoned terraces are subject to sudden and massive increases in erosion of their soils. This is the effect when terrace walls break down, and gully erosion begins to speed up the runoff of water.

There are a variety of other structures which belong to the Andean technology. Related to the raised fields there are narrow ridges, used for growing potatoes and vegetables generally in intensive farming, and termed *tarasukas* or *tara sukas or sukallas*, in Aymara. These are valuable on poorly drained lands, where they work through imitation of the branching network of a stream, taking water off efficiently and allowing cultivation on the ridges between. Another feature in poorly drained land, seen in both Bolivia and Peru, are the *huisu kollu*, double-raised beds, in other words short raised beds standing on top and at right angles to the main, larger raised fields beneath. In areas of limited water supply such as the area 30 km. north of Juliaca, and the Copacabana peninsula in Bolivia, there are systems of ponds, the *qochas*, some of them interconnected by canals, where the pond bottom is cultivated as the dry season proceeds. These remain in use, though unstudied and never adopted as a technology by the Spaniards or modern farming. In drier parts of the altiplano too, especially in Bolivia, there are local irrigation systems, fed by springs or streams in the mountains, and linked through storage tanks along the way, the qotañas, which help stabilize the flow. Sunken fields, most studied along the dry Pacific coast of Peru, are also known in the altiplano, dug down to the water table so as to be able to harvest the water available there. A feature shown to me by Eduardo Chilon in the altiplano near Batallas on the altiplano of northern Bolivia, is what Chilon[1] calls "water harvesting" on the hill surfaces, building of low earth walls down hill slopes where there is an amphitheatre shape to focus the flow of surface water and collect it at the base in fields which have intensive cultivation, or which serve as permanent watermeadows for animals when droughts occur.

Of course, some parts of the Andean technology are not engineering structures at all, but the use of particular crops and crop combinations, or the timing in use of fields and their rest periods. One notable example is the use of a well-established fallow system, under which the community agrees to maintain their cultivated land, even when individually owned, in several fields or *aynokas*, the use of each field being agreed in advance of the growing season and for many seasons to come. These *aynokas* are rotated in use in what is termed a "sectoral fallow system" (Pestalozzi 2000, Morlon 1992).

1 Eduardo Chilon, Escuela Militar de Ingenieria, La Paz, unpublished document.

There may be as many as 12 or 15 fields, which allows each field to be used in a three or four year rotation of crops, commonly potatoes, followed by quinua and then oats or barley. After the crop rotation, there is a fallow of eight, nine or ten years, which allows the field to recover its nutrients, but more importantly, to build up biomass in the root systems of field weeds, which will be liberated when the field is cultivated again. This system is deceptive, since the fields look abandoned for many years, and the poor surface plant growth of weeds in the fallow fields suggests a totally non-productive interval. In fact, as Pestalozzi's research group found out for a high (over 4000 metres) region of Cochabamba province in Bolivia, the fallow period serves to build biomass below the surface, in the root systems which hold over 90 per cent of the biomass total. When cultivation recommences, the response of crops such as potatoes is immediate and strong.

As recorded in Chapter two, all these systems fell into total or partial disuse before colonial time, but they are likely to have operated jointly in a highly complex system of land management in Tiwanaku times. Whatever the reasons for abandonment, it was total and all-encompassing, so that the linkages to other parts of the system were also forgotten.

Integration

Listing these different elements of the Andean technological armoury has the further inference: projects which attempt to bring back to life just one element are unlikely to succeed, since they are placed in a new context and must compete with other alien systems. It is important to discuss the matter of integration, even if we are unsure of the ways in which linkage between one element and another took place. Some of the forms of integration are in fact fairly well known. We have mentioned above, the linkages between terraces on the slope land, and raised fields on the plains, cited by Donkin (1979). His concept can be extended. Raised fields for drainage, close to the Lake, were complemented by other raised fields further back, whose function was more frost avoidance than drainage; above these were the terraces, rising from the altiplano plains level at 3800-3850 metres, to about 4200 metres, where the more tender crops could be grown.

The kind of integration which has been most fully discussed is the vertical linking of different altitudes over a wide range, from tropics to altiplano, which has been identified by John Murra (1975) from historical documents, and separately by Ramiro Condarco (Condarco and Murra 1987). The idea of a combined use of many ecological niches, from the lower Yungas where tropical fruits or coca might be produced, up to the highest levels where only the camelids can prosper on coarse grasses, is attractive in itself, and is documented historically as well as today. Although it cannot provide us with a basis for the explanation of Inca sustenance, it remains as one aspect of integration, requiring complex relationships and organization of production to operate at all. Labour time, land ownership, and social relations needed to be established as well as trade routes and marketing arrangements for the bringing of products from distant valleys into the heartland of the altiplano.

Today, this kind of integration still exists for many village communities of the Central Andes. An example is the community of Coscapa, near La Paz, where vertical integration of work and land use is still found, despite the gradual breakdown due to the proximity of La Paz. Here at 3500-4500m on the east side of the Cordillera Oriental, in a relatively humid environment, three altitudinal levels are generally recognised, each with several subdivisions:

1) The high level or Puna, where land uses include cultivation of the bitter potato and the breeding of sheep and llamas. The ancient three field system, of *aynokas*, is used for organizing production and all farm families have access to the land for grazing and cultivation.
2) The middle section, or valley heads, has the settlement of Coscapa itself, at around 3400 metres, and extends down to 2200 m. This section has most of the cultivation, for common potatoes, and the standard subsistence crops of maize, peas and beans. There are many terraces in use, some with irrigation, for relatively intensive cultivation. Cropland is mixed with pastures for cattle sheep and camelids.
3) There is a lower section down to the River Zongo, below 2200 m. where some individuals have inherited or assumed land rights. Hard maize for cattle feed, corn on the cob, squash, locoto, racacha and walusa are grown, and some beef and dairy cattle bred, though the heat brings problems of sanitation for cattle.

On the altiplano itself, "horizontal" access to varied resources is also important. Donkin's integration of raised fields with the terraces has been mentioned, but the linkages can be varied and in connection with a variety of climatic and other factors. Different distances from the lake give different temperature regimes – more extreme climates are found with greater distance from the lake as the moderating influence of a great water surface are lost. On the altiplano, the soils vary, with poorly drained but highly fertile soils near the lake edge, thin soils with a calcareous hardpan over many higher surfaces, and irregular depressions or patches along old river courses with deeper soils. The mountains extending up above the altiplano give totally different environments, with extensive grazing lands up to the limit of occupation near 4500 metres. In the highlands, small hollows with deep black soils, the *bofedales*, provide especially rich pastures which do not dry out in the dry seasons. Integration of livestock management with cultivation is a key element for success (its lack has been mentioned as a factor in failure locally), and it must have been achieved by ancient communities. They would have kept cattle on higher grounds above the plain, and on resting fallow land on the plain. Cultivation would have been on chosen sites and carefully protected. Today, this integration seems lacking, but it would be of equal importance to achieve, simply because much of the land is in fact oriented more to livestock than to crops. Certainly, the whole of the Bolivian sector of concern, as well as all the higher land (above about 3900 metres) in Peru, must be regarded as cattle country rather than land for cultivation.

Often the horizontal access to different lands may be important because of accidents of weather or hydrology. Frequent hailstorms attack the altiplano in some growing seasons, and these may hit one small patch of land, and leave another, two kilometres away, unscathed. A river may flood (the altiplano has slight slopes and rivers with heavy silt load, liable to flooding) and leave one settlement's crop land destroyed, its neighbour unaffected.

In these kinds of case, communities with large land areas had an advantage. They could move cattle out of flooded land, replant on lands unaffected by the accident, whether on communal land or that owned by other members of the same community. This kind of access is of course being destroyed by the movement to private individual holdings favoured by the current agrarian reforms. The area around Acora, to the south of Puno, provides an example. Here, most communities hold some hill land above the altiplano and up to 4300 metres, for grazing their animals, while their main production base is the drained plain around the town of Acora itself. In addition, a number of them have frontages onto Lake Titicaca, where new fertile land is exposed whenever the lake retreats, and which has a privileged climatic regime.

In an area north of Puno, around Huatta, there is a block of community land but no nearby hills of any height. Here however, the elders of the community reported that they could always have access to higher land, by way of exchange, as relatives might agree to take their animals, in return for grain or tubers as a food source, when there was a need. Horizontality relies thus, not just on physical proximity, but also on social linkages.

Andean Cosmovision

Beyond all physical integration, there was also a spiritually holistic view of the world in the Central Andes. Their world view or "cosmovision", as it is now increasingly being termed, was of a three part cosmos, sky, land surface, and inner earth, in which mankind exists and forms another part, with no special priority or privileged position, in contrast to the Judaic and Muslim traditions, which place mankind as a separate kind of being, living with Nature but often in conflict with it. In this system, with elements of animism i.e. the belief that all species in nature, and many special places, are endowed with some magic or mystical properties, the attitudes of man to nature are necessarily different from our own. Christian and similar religions have inspired the belief that we can act to dominate our environment, and have every right to use its resources without hindrance. A cosmovision that makes us just one amongst myriad species, by contrast, means that we must act with care and circumspection, to work with other elements of Nature to achieve results that help us but do not upset the rest of life. Integration, the topic with which we are dealing, is something that occurs naturally if we believe that we are as a species, only one part in an ecological whole.

Some aspects of this *Weltanschauung* are translated into action through the mediation of a series of rituals or ceremonies associated with the main events of the agricultural year – turning the soil in January, cleaning tanks and canals in June or July, providing a watch over crops through the growing seasons, all these are associated with local ceremonies in the fields or the villages. This Andean agroecology is discussed in Chapter six. Ritualization is of course highly functional in human terms too; in communities where there is still communal land and communal work, there is a need for organizing the workforce to operate together, and for there to be agreements over the management of land for the benefit of all. To mark this organization with a ritual endows the operation with some special prestige. In the communities studied in detail, there is no strong linking of raised fields with the various ancient ceremonies, which when practised are limited to the areas of long cultivation, the *ladera* lands around the village houses.

The Raised Fields, Diversification and Integration

It is worth asking specifically now, what role is held by the raised fields within the system of diversified farming described for the Central Andes. As a small area in any one community, one or two hectares usually, the technology is obviously not used at the expense of all the other traditional elements. The kinds of analysis discussed above have the idea that its main purpose may be risk minimization, which is the accepted explanation of how peasant farming works. But this technology is not really an attempt at risk minimization. For all farmers in the region, to build the raised fields has been a risk in itself, an investment of many manhours in consultation, construction, and maintenance. These people are not risk avoiders, but risk takers, as there is no guarantee that the fields will provide a good return, and most have not yet proved their worth. As Kervyn (1992) points out, Central Andean farmers can be regarded as like modern Western farmers in many respects, willing to try out new crops and techniques, but also seeking to limit the damage caused by their excessively variable environment.

What raised fields provide is another element in the technological package, and one which can come into play as a compensatory factor when natural catastrophe threatens. In the case of flooding, they offer the prospect of a crop when most of the land has been flooded, as occurred in 1985-86 and again in 2000-2001. In the case of drought, they offer a crop at least on lands near the lake or in hollows, where groundwater may be tapped and where the limited rainwater may be accumulated for longer than on the flat pampas.

Clark Erickson emphasizes this point in historical and prehistorical terms, showing that farming has always been in fluctuation in the lake side lands over the last 3000 years, but that the farmers and their organizations have managed to survive due to their use of, among other things, raised

5. Shepherd girl and sisters at Chucuito, south of Puno, watching a lamb in the totora reed beds by Lake Titicaca, while their father is out fishing. Livestock and fishing are important elements in the farming system. Diversity and using alternative resources are fundamental to these people's economic strategy.

6. Lima Pampa, a rural settlement north of Juliaca, Peru. On this treeless plain there are no good fuel sources and the animal dung is dried on the adobe walls around the village huts, to be burnt in open fires or stoves for cooking. Farming here is more geared to livestock than arable, which might have had a better use of the dung for manure.

fields in the newly exposed lake edge rich soils when long droughts have occurred, and raised fields further away from the lake at times of flood.

Today, the same reasoning holds good. Farmers in interview, commented to me that the raised fields were something to be held in readiness, to be used when their time came, that they were well understood and would always be appreciated. A first conclusion must be that the whole technology has been absorbed, and will be used from time to time in a system characterized by diversity of techniques, of land areas used, and of products.

On the other hand, the raised fields are not well *integrated* into the farming system. Within the farming year, there is no time which is particularly set aside for work on the building or maintenance of the structures, and time has always to be borrowed for new construction or for any work which is needed, from other jobs around the farm. Farming intensively on the raised fields is also not integrated with the raising of livestock on the land. Traditionally, livestock have been regarded as well integrated with the cultivation of the land. They provide dung for the fields, grazing over stubble after the harvest, and breaking hard dried clods of earth. They also use the high puna lands on steep slopes which would otherwise be unusable for cropping; and they provide power for working the land. But in effect, in the Lake Titicaca area, as elsewhere in the Central Andes (McCorkle 1987), there are a number of conflicts to be seen. The livestock are still grazed over the whole plains area, and there are no hedges, stone walls, or fences, over 90 per cent of the land area. This means that careful shepherding is needed to keep animals out, especially from areas growing succulent and attractive crops. Shepherds are no longer easy to find and this control activity is in consequence deficient. Additionally, the land has been subdivided or parcelled up, since the 1980s, and the communal land which was used for grazing in fallow years, and where communal control of the grazing was possible, is no longer available. The livestock lack fodder, they lack access to different lands such as the *bofedales*.

The central problem for raised fields resulting from the lack of animal integration, is the maintenance of fertility in the soil. We do not know how fertility was maintained in ancient times, or whether it was restored after exhaustion, by long fallows. But the present purpose and aim of the programmes, to have permanent intensive use of the fields, is frustrated by the decline of fertility in the land, resulting in abandonment after four or five years. This could be overcome by heavy use of animal dung, but there is no system for incorporating this dung on the land at the present. What the agencies have proposed, the use of alfalfa to bring back nitrogen to the soil in a rotation of several years followed by renewed cropping with potatoes and the other crops, has not yet been proven as a system. On the other hand, local farmers do not regard fallowing the raised fields as a problem, but an acceptable way of using the land. Given the irregular value of the fields in their system, at times of flood or drought, it is perfectly logical to have these lands as a reserve. In fact, to put more effort into the raised fields would be a misuse of time, as they might not be of great value in normal years.

What Next?

We arrive at the overall conclusion that there is a mismatch between the aims of agencies and institutions, and those of the small farmers of the altiplano. Failure for the first group is success or at least something acceptable for the second. Diversification is always acceptable to subsistence farmers, as it helps ensure survival.

It would have been most desirable for the raised fields to have been brought back as part of a total farming system in the lakeside area. But to achieve the aim of an integration between the raised fields and the rest of the current Andean farming system, would need something far beyond what has been attempted so far. That is, a massive programme of rural development, which attacks all aspects of farming life together, bringing a variety of crops into the farm, encouraging all the different kinds of Andean technology (terraces and ponds as well as raised fields), together with a real attack on soil problems such as salinity and declining fertility. Such a programme would also tackle the vertical chain of production, looking at transport systems to get inputs onto the farm and production from the farm to store houses and to market. Such ideas have of course been around for a long time, in the writings on integrated rural development from the 1980s. The experiments around Lake Titicaca could be classified as going back to before that time, to the efforts at single theme projects of development common in the 1970s or earlier, throughout the poor tropical countries. In the long term, and assuming that the efforts to maintain traditional culture and land use in the Andes are finally a failure, there is the possibility that in fact raised fields could be successful with a more capitalistic type of farming. Though the fields are small, they could be used by mini-tractors, and the raising of new fields could be done with earthmoving machinery, if this was sophisticated enough to avoid turning up subsoils.

On the ground, the absence of what the archaeologists and others had dreamed of – a landscape of intensive cultivation with dense population and a highly productive agriculture – must be accepted. It may well be that raised fields, with associated technologies, were the reason why a high culture could have been created on the high plateau, and that this system has been truly rediscovered by modern science. It is likely too, that it was a sustainable land use involving only locally available resources, and was highly successful over many centuries. But its reincorporation into local farming is of a different nature. For the near future, the use of raised fields is likely to be as an occasional alternative to other land uses, at long intervals with rest periods in between, and at times when the communities feel threatened by flood, frost or other hazards.

Bibliography

Achata, Adolfo, and Jesus Sanchez (1990), *Evidencias de Cambios Tecnologicos y organizativos y sus impactos en la agricultura de dos comunidades campesinas de Puno*, INIAA-PISA, Serie Tecnica, Documento de Trabajo no. 1, Puno, Peru.

Albarracin-Jordan, Juan (1999), *The Archaeology of Tiwanaku: the Myths, History and Science of an Ancient Andean Civilization* (published privately) La Paz.

Allen, J.M. (1998), *Atlantis: the Andes Solution*, The Windrush Press.

Altieri, Miguel A. (1995), *Agroecology: the science of sustainable agriculture*, Westview, Colorado.

Altieri, Miguel A. (2002), "Agroecological principles for sustainable agriculture", Ch. 3, pp. 40-46, in Uphoff, Norman (ed).

Avila Cazorla, Edgar (2002), *Cadena productiva de vacunos de carne*, Ministerio de Agricultura, Direccion Regional Puno.

Baudin, Louis (1955), *El imperio socialista de los Inca*, Zig-Zag, Santiago; also in translation as Baudin (1961), *A socialist empire; the Incas of Peru*, D. Van Nostrand, New York.

Bellamy, H. and Allen, P. (1948), *The calendar of Tiahuanaco*, Faber and Faber, London.

Bennett, Wendell Clark (1934), "Excavations at Tiahuanaco", *Anthropological Papers of the American Museum of Natural History* Vol. 34/3, pp. 359-491.

Bennett, Wendell Clark (1935), "Excavations in Bolivia", *Anthropological Papers of the American Museum of Natural History*, Vol. 35/4, pp. 329-507.

Berastain, Juan Palao (ed) (1992), *Experiencias en rehabilitacion de Waru Waru; Puno 1981-89*, PIWA/INADE, Puno.

Berastain, Juan Palao (1999), *Waru Waru: cultura y desarrollo en el altiplano del Titicaca*, PIWA/INADE, Puno.

Boserup, Ester (1955), *The conditions of agricultural growth; the economics of population pressure*, Aldine Press, Chicago.

Browman, David L. (1987), *Arid Land Use Strategies and Risk Management in the Andes*, Westview, Boulder, Colorado.

Bryson, R.A. (1978), "Cultural, economic and climatic records", Ch. 7.3, pp. 316-327, in A. B. Pittock, L.A. Frakes, D. Jenssen, J.A. Peterson, J.W. Zillman (eds) *Climatic Change and Variability*, Cambridge University Press, Cambridge.

Chire, Hector and Valdivia, Marco (2000), *Evaluacion agroeconomica comparativa de la produccion agricola en waru waru y pampa en un ciclo productivo*, PIWA/PELT, Puno.

Condarco, Ramiro (1970), *El escenario andino y el hombre: ecologia y antropogeografia de los Andes Centrales*, La Paz.

Condarco, Ramiro, and John Murra (1987), *La teoría de la complementariedad vertical eco-simbiótica*, Hisbol, La Paz.

COSUDE (Cooperacion Suiza para el Desarrollo) (2000), *PROSUKO - Informe Anual de Actividades 1999-2000*, manuscript at COSUDE offices, La Paz, Bolivia.

COSUDE (2000A), *PROSUKO; Plan Operativo 2000-2001*, La Paz, Bolivia.

COSUDE (2001), *La tecnologia de los Suka Kollus desde una perspectiva social*, La Paz, Bolivia.

Cotterell, Maurice (2001), *The Lost tomb of Viracocha: unlocking the secrets of the Peruvian pyramids*, Hodder Headline, London.

Daniken, Eric von (1968), *Chariots of the Gods*, Bantam Books, New York.

Delgado, Freddy (ed), *Cosmovision indigena y biodiversidad en America Latina*, Compas/AGRUCO, Cochabamba.

Denevan, William (1966), *The aboriginal cultural geography of the Llanos de Mojos of Bolivia*, Ibero-Americana 48, UCAL Press, Berkeley.

Denevan, William (1970), "Aboriginal drained-field cultivation in the Americas", *Science*, Vol. 169, pp. 647-54.

Denevan, William (1986), *The cultural ecology, archaeology and history of terracing and terrace abandonment in the Colca Valley of Peru, Vol. 1*, Technical Report to the National Science Foundation, Dept. of Geography, University of Wisconsin, Madison, Wisconsin.

Denevan, William (2001), *Cultivated landscapes of native Amazonia and the Andes*, Oxford University Press, London.

Diaz, C. and E. Velasquez (1990), *Mapa fisico de camellones o Waru Waru*, PELT-INADE, Puno.

Donkin, R.A. (1979), *Agricultural Terracing in the Aboriginal New World*, Univ. of Arizona Press, Tucson.

Erickson, Clark L. (1988), *Raised field agriculture in the Lake Titicaca Basin; putting ancient agriculture back to work*, University of Pennsylvania Museum, Pennsylvania, USA.

Erickson, Clark L. (1996), *Investigación arqueológica del sistema agrícola de los camellones en la cuenca del lago Titicaca del Peru*, PIWA, Puno.

Erickson, Clark L. (1999), "Neo-environmental determinism and agrarian 'collapse' in Andean prehistory", *Antiquity* 73, pp. 634-42.

Erickson, Clark L. and Kay L. Candler (1989), "Raised fields and sustainable agriculture in the Lake Titicaca Basin of Peru", Ch. 14, pp. 230-50, in Browder, John O., *Fragile Lands of Latin America: strategies for sustainable development*, Westview, Boulder and London.

Espinoza Soriano, Waldemar (1997), (3rd edn.), *Los Incas: economia, sociedad y estado en la era del Tahuantinsuyo*, Amaru Editores, Lima.

Flores Ochoa, Jorge A. (1987), "Cultivation in the Qocha of the South Andean Puna", Ch. 12, pp. 271-296, in Browman (ed).

Gourou, Pierre (1966), *The Tropical World: Its Social and Economic Conditions and its Future Status*, Longmans, Green & Co Ltd, London.

Graffam, Gray (1992), "Beyond state collapse: rural history, raised fields, and pastoralism in the South Andes", *American Anthropologist*, Vol. 94/4, pp. 882-904.

Hamilton, Roger (2000), "Soldiers present arms - and turnips", *IDB America*, July-August 2000, pp. 18-19, Interamerican Development Bank, Washington, DC.

Hastorf, Christine A. (1993), *Agriculture and the onset of political inequality before the Inka*, Cambridge U.P., Cambridge.

Inbar, Moshe and Llerena, Carlos A. (2000), "Erosion processes in High Mountain agricultural terraces in Peru", *Mountain Research and Development*, Vol. 20/1, pp. 72-79.

Intercooperation (1992), *Estudio de factibilidad del proyecto Suka Kollu*, Berne, Switzerland.

Izquierdo Condori (1992), "*Analisis economico de la produccion agricola en waru waru. Comunidad campesina de Cochela 1990-91*", in PIWA 1992a, pp. 113-139.

Kervyn, B. (1992), 'L'économie paysanne au Pérou: théories et politiques', Ch. 7. pp. 437-470, in Morlon, Pierre (1992).

Kolata, Alan (1982), "Tiwanaku: portrait of an Andean civilization", *Bulletin, Field Museum of Natural History, Chicago*, September, pp. 13-28.

Kolata, Alan (1993), *The Tiwanaku: portrait of an Andean civilization*, Blackwell, Oxford.

Kolata, Alan (1996), *Valley of the Spirits: a journey into the lost realm of the Aymara*, Wiley, New York.

Kolata, Alan L. and Charles Ortloff (1989), "Thermal analysis of Tiwanaku raised field systems in the Lake Titicaca Basin of Bolivia", *Journal of Archaeological Science* Vol. 16, pp. 233-63.

Lennon, T.J. (1983), "Pattern analysis of pre-hispanic raised fields of Lake Titicaca, Peru", pp. 183-200, in Darch, J.P. *Drained field agriculture in Central and South America*, BAR International Series 189, BAR, Oxford.

McCorkle, Constance M. (1987), "Punas, pastures and fields: grazing strategies and the agropastoral dialectic in an indigenous Andean Community", chapter 2, pp. 57-79, in Browman, D.

Morlon, P. (1981), "Questions sur l'agriculture de l'altiplano peruvien", in *L'homme et son environnement a haute altitude*, CNRS, Paris, pp. 107-114.

Morlon, P. (ed) (1992), *Comprendre la agriculture paysanne dans les Andes Centrales: Perou-Bolivie*, INRA Editions, Paris.

Morris, Arthur (1981), *Latin America: economic development and regional differentiation*, Hutchinson, London.

Morris, Arthur (1999), "The agricultural base of the pre-Incan civilizations", *Geographical Journal*, Vol. 163/3, pp. 286-95.

Morris, Arthur (2001), "Land reform and technology in Puno, Peru", Chapter 8, pp. 125-38, in Zoomers (ed).

Murra, John V. (1975), "El control vertical de un maximo de pisos ecologicos en la economia de las sociedades andinas", Chapter 3, pp. 59-116, in *Formaciones economicas y politicas del mundo andino*, Inst. De Estudios Peruanos, Lima.

Ortloff, C. (1989), "Analisis termometrico de los sistemas de campos elevados en el area de la Pampa Koani", pp. 237-61, in Kolata, A., *La tecnologia y organizacion de la produccion agricola en el estado de Tiwanaku*, Inst. Nacional de Arqueologia, La Paz.

PELT/COSUDE (2001), *Informe de Avance del Proyecto PIWA*, Piwa, Puno.

Pestalozzi, Hansuell (2000), "Sectoral Fallow systems and the management of soil fertility: the rationality of indigenous knowledge in the High Andes of Bolivia", *Mountain Research and Development*, Vol. 20/1, pp. 64-71.

PIWA (1992), *Testimonios campesinos; experiencias sobre (re)construccion y produccion agricola en waru waru, en las comunidades campesinas de Caritamaya, Coata y Chatuma*, Convenio PELT/INADE, Puno, Peru.

PIWA (1992A), *Avances de investigacion sobre la tecnologia de waru waru: III: socio economia* Convenio PELT-INADE, Puno.

PIWA (1994), *Microclimatologia en el agroecosistema de waru waru; avances de investigacion*, Convenio PELT-INADE, Puno.

Plafker, George (1963), "Observations on archaeological remains in Northeast Bolivia", *American Antiquity*, Vol 28, pp. 372-78.

Polanyi, Karl, Arensberg A. and Pearson, Harry W. (1957), *Trade and markets in the Early Empires*, Free Press, Glencoe, Illinois.

Ponce Sangines, Carlos (1994), *Arthur Posnansky y su obsesion milenaria: biografia intelectual de un pionero*, CIMA, La Paz.

Posnansky, Arthur (1945), *Tihuanacu; the cradle of American man*, New York, J.J. Augustin.

PROSUKO (1995), *Areas potenciales para la introduccion de Suka Kollus*, internal document, PROSUKO, La Paz.

Quispe, Jesus, Diaz, Marina, and Peralta, Alcides (1999), *Perspectiva de genero en la extension de la tecnologia waru waru*, Delgado, Puno, Peru.

Rocha, R., Vargas, A. Saavedra, R. Marin, P, Yuca, E. (2000), *La Tecnologia de los Suka Kollus desde una perspectiva social*, Cosude, La Paz (internal document).

Rojas, Severo (2001), "Crianza ritual de la papa", pp. 355-360, in Delgado (ed).

Rondinelli, Dennis A. (1985) *Applied Methods of Regional Analysis: The Spatial Dimensions of Development Policy*, Westview Press, Boulder and London.

Ruddell, Edward D. (2002), "Raising smallholder crop and livestock production in Andean Mountain Regions", Ch. 14, pp. 173-186, in Uphoff, *op cit.*

Ryder, R. and Knapp, G. (1983), "Aspects of the origin, morphology and function of ridged fields", in Darch, J. P. *Drained field agriculture in Central and South America*, BAR, Oxford, pp. 201-220.

Salas, Porfirio, Peralta, Alcides, and Mollocondo, Charles (2000), *Evaluacion del impacto socioeconomico de los waru waru en familias del ambito PIWA III fase (1995-98)*, PIWA, Puno.

Sauer, C. (1974), *Agricultural Origins and Dispersals*, in *Seeds, Spades, Hearths, and Herds*, 2nd Edition, pp. 1-110, MIT Press, Cambridge, Mass.

Schulte, Michael (1996), *Tecnologia Agricola Altoandina; el manejo de la diversidad ecologica en el Valle de Charazani*, Plural Editores, La Paz, Bolivia.

Shimada, I., C.B. Schaap, L.G. Thompson and E. Moseley Thompson (1991), "Cultural impacts of severe droughts in the prehistoric Andes: applications of a 1500 year ice-core precipitation record", *World Archaeology*, Vol. 22 pp. 247-270.

Smith, Clifford T., Denevan, W.M. and Hamilton, P. (1968), "Ancient ridged fields in the region of Lake Titicaca", *Geographical Journal*, Vol. 134/3, pp. 353-67.

Smith, R. (1987), "Indigenous agriculture in the Americas; origins, techniques and contemporary relevance", in Preston D. (ed) *Latin American Development; geographical perspectives*, Longmans, Harlow.

Squier, Ephraim George (1877), *Peru: incidents of travel and exploration in the land of the Incas*, Harper, New York.

Stubel, Alphons and Max Uhle (1892), *Die Ruinenstatte von Tiahuanaco im Hochlande den Alten Peru. Eine Kulturgeschichtliche Studie*, Verlag von Karl Hiersemann, Leipzig.

Thorner, Daniel, Kerblay, Basile, and Smith, R.E.F. (eds) (1966), *A. V. Chayanov on the theory of peasant economy*, Irwin, Homewood, Illinois.

Uphoff, Norman T. (2002), (ed), *Agroecological innovations; increasing food production with participatory development*, Earthscan, London.

Ventura Lope, Vicente (1992), "Lo que piensa la gente en la comunidad campesina de Caritamaya-Acora", pp. 11-42, in PIWA (1992).

Villaroel, Tito, David Salazar and Raul Delgado (2001), "La biodiversidad de los tuberculos andinos; importancia en la vida de las familias de una comunidad altoandina de Cochabamba", pp. 295-308, in Freddy Delgado (ed) (2001) *op. cit.*

Villena Soria, Jaime (1996), *Efectos de los incentivos en las posibilidades de adopcion de la tecnologia waru waru*, PIWA, Puno.

Zeballos, Cesar Diaz (1994), *Priorizacion de las areas potenciales para la reconstruccion de waru waru en el altiplano de Puno*, INADE/PELT COTESU, Puno.

Zoomers, Annelies (ed) (2001), *Land and sustainable livelihood in Latin America*, Royal Tropical Institute/KIT Publishers, Amsterdam, Holland.

Index

abandonment, field 3,18, 30-32, 34, 53, 87, 93, 103, 105, 107, 109, 116, 125
Achuta Grande 86-87
Acora 22, 73-74, 92, 111, 125
adobe 37, 78, 81, 92, 115
  blocks 74
aerial photography 1, 16, 42, 44
agrarian reform 53, 59, 67, 81-82, 86, 90, 111
air 4, 19, 67, 69, 73-74, 91, 107
  drainage 107
Albarracin-Jordan, Juan V. 6, 11, 26, 28, 31, 33, 125
alfalfa 45, 73, 77, 82, 93, 117
  seed 46, 80
Ancoraimes 50, 84-87
andean technology 18, 27, 35, 63, 104, 106, 108, 117
animals 28, 37, 41, 62, 64, 72-73,77, 80, 82-84, 86, 88, 92-93, 99, 108, 111-112, 114-116, 123
  animal husbandry 77, 87
  cattle 48, 73-74, 78, 82-83, 87, 110-111
  sheep 73-74, 83, 86, 98, 110
Arequipa 74
Aricollo 83, 87
asistencialismo 46, 66
Aygachi 17, 53
Ayllus 90
Aymara 1-2, 7, 10, 25, 31, 34, 40, 47-49, 69, 92, 103, 108, 122
  language 40
  nationalist movement 40
  Señorios, lordships 32
ayni 54, 98
aynokas 75-76, 92-93, 97, 108, 110
  see also open fields; three field system

barley 20, 39, 48, 72-74, 77, 81-84, 86-87, 92, 107, 109
  rust 95
Batallas 48, 50, 53-54, 69, 108
Baudin, Louis 11, 16, 119
biomass 91-93, 109
Bofedales 111, 116
Bolivian 47
  archaeology 10, 13, 15, 17, 25, 29, 33
  national parks 9
  nationalism 7, 10, 14
  territory 8
Boserup, Ester 28, 105, 125
broad beans 48, 86, 93
Buddleia incana 9, 12, 84

Canahua, Alipio 47, 63, 79, 94
cañihua 82, 92
CARE International (The Cooperative for American Remittances to Europe – formed in 1945) 46-47, 53, 63, 69, 71, 73, 76-79, 82-83, 93-94
Caritamaya 20, 73, 75-79, 123-124
Catari, river 26
  valley 31
Catholic *see* University
chakitaclla 57, 73

Chayanov, A. 61, 106, 124
Chura, Ernesto 82
climate 5-7, 9, 12, 17, 19-20, 30, 32-33, 46, 58-59, 68, 105, 111 *see also* microclimate; temperature range
club de madres 1, 65, 77, 79-81
  Doña Gladys (president) 79-81
comunidades campesinas 119, 123
cosmovision 112, 120
  Andean 112
cost-benefit analysis 2, 34, 61-62
COSUDE (Cooperacion Suiza para el Desarrollo) 19, 45, 49-50, 53-54, 69, 72, 120, 122-123
crianza 98, 123
Cuzco 6, 25, 30

Denevan, William 14, 16-17, 35, 42, 44, 106-107, 120, 124
discourse 66, 70
diversification 18, 40, 53, 62, 80, 86, 90-91, 94-95, 102, 105-107, 113, 116
Donkin, R. 107, 109-110, 120,
drought 30, 33, 35-36, 58, 66, 69, 75, 104, 108, 11, 116-117, 124 *see also* rainfall deficit

ecological niches 95, 110
economy 47, 72, 74, 90, 98, 103, 106, 124
  capitalist 61
  EMI *see* University
  subsistence 61, 72
El Niño 35
Erickson, Clark 13-15, 18-19, 24, 26-30, 32-35, 41-42, 47, 63, 105, 113, 121

fallows 108, 118
  long term 116

floods 1-2, 16-18, 22, 26-27, 33, 35, 58-59, 62, 66, 69, 75, 77-78, 80, 103-105, 107, 111, 113, 116-118
fodder 86-87, 116
FONCODES (Fondo de Compensacion de Desarrollo Social) 82 *see also* state agencies
frosts 2, 19-20, 48, 55, 58, 62, 66-67, 69, 75, 81, 84, 91, 99, 103, 109, 118

genetic drift 95
genetic erosion 90, 94-95
gorgojo de los Andes 100 *see also* pests
Graffam, G. 32-34, 121
Green Revolution 37, 40
greenhouses 37, 84
  plastic roof 37

haciendas 5, 73
  Buenavista 66, 67
harvest 5, 39, 53, 57, 61-62, 66, 72-73, 77, 80, 97-99, 108, 116
horizontal integration *see* integration
Huatta 16, 19, 27, 32, 38-39, 41-43, 47, 66, 92, 100-112
huertos familiares 37
huisu kollu 85, 108
hydraulic civilization 24, 29

Illpa 42
  Agricultural Experimental Station 63
  river 58, 63, 81
Inca 4, 6, 10, 13, 24-25, 28, 31-34, 63, 110, 119, 121-122, 124
indigenous knowledge 35, 40, 122
integration
  animal 96, 116

farming systems 97, 102, 106, 109-113, 17
horizontal 107, 110-111
spatial 70, 96
temporal 70, 96
vertical 69, 109-110, 117, 120, 122
irrigation 18, 20, 63, 81, 110
canals 4, 6, 16, 22, 25-27, 55, 57, 63, 81-82, 85
systems 25, 29, 34, 91, 93, 100, 107-108

joqho 69
Juliaca 13, 23, 41-42, 58, 63, 74, 78, 92, 108, 115

karañas 69
kiswara, see Buddleia incana
Kolata, Alan 11, 15, 17-19, 24-26, 29-35, 44, 47-50, 84, 104, 121-12
Kollanas 9
kollus 1, 53, 55, 86
kurakas 25

labour 5, 14, 24-27, 41, 54, 61-63, 67, 72-73, 80, 82, 86, 96, 104-106
forced 24
slavery 5, 16, 24, 59
laderas 58
land tenure 40, 59, 63, 66, 77, 80-81, 96, 104
laymi 92
Lima 7, 45, 59, 63, 74, 115, 121-122

maca 94
machines 39-40, 55, 57, 62, 68, 73-74
earthmoving 118
heavy 89
maize 12, 34, 48, 99, 110
manure
animal 80, 92, 105, 115
vegetable 105
marketing 46, 53, 9, 74, 77, 110
markets
commercial 5, 46, 74, 83, 90, 94, 123
Sunday 99
mechanization 55, 61, 63, 69
animal traction 73
oxen 65, 73, 98
tractors 55, 62-63, 68-69, 77, 82, 97, 118
microclimate 35, 85, 91 *see also* climate; temperature range
migration 12, 65, 67, 72, 98
minka 54, 97, 98
mit'a 25
Mitma 34
Mojos, plains of 14, 16, 120
Mothers Clubs, *see club de madres*
Murra, John 109-110, 120, 122

National Revolution (Bolivia) 10, 59
NGO (Non-Governmental Organisation) 42, 48-50, 68, 84, 95, 103

olivillo 12
onions 48, 74, 83, 98
open fields 75, 76 *see also aynokas*

Pacajes 28, 33
Pachamama 98-99
Pampa Koani 11, 19, 24, 26, 29, 31, 34, 48, 122
Paucarcolla 58, 78-81, 87
Paucarmayu 81, 82
pests 65, 89, 100 *see also gorgojo de los Andes*
phalaris 84

PIWA (Programa Interinstitucional de Waru Waru) 3, 20, 22, 37-38, 45-47, 50, 58-60, 69, 71, 76-77, 79-83, 119-124 *see also* state agencies
Polylepis incana *see queñua*
Ponce Sangines, Carlos 7, 9-10, 123
ponds 13, 31, 78, 108 *see also qochas; qotas*
population 2, 4-5, 7, 11, 18, 24, 26, 32, 46, 72-73, 87, 102, 106
  pressure 67, 72, 93, 119
  rural 67, 72, 103, 105
poruma 69
Posnansky, Arthur 7-11, 15, 30, 123
potatoes 1, 20, 38, 48, 73-74, 77, 79-82, 84, 86, 92, 98-99, 107-110, 117
  bitter potatoes, Ruckii species 20, 38
  seed 76, 80
  Solanum tuberosum 20
Potosi 95
PROIMPA 49 *see also* NGO; state agencies
PROSUKO (Programa Interinstitucional de Suka Kollus) 19, 49-54, 65, 67, 72, 85-86, 93, 120, 123 *see also* state agencies
Pukara 13-14, 28
Puno 16, 20, 37, 39, 41-42, 45-47, 58, 63, 67-68, 73, 75, 78, 80, 83, 94, 97, 111-112, 114, 119, 120-124
  Departamento 37

qochas 13, 29, 31, 33, 73, 108
qotañas see qochas; qotas
qotas 77 *see also* ponds
Quechua 1, 2, 25, 47, 92, 103
queñua 12
quinua 20, 68, 73, 77, 81, 86, 94, 109
quishuara see Buddleia incana

rainfall deficit 23, 31 *see also* drought
raised fields *see waru waru*
reciprocity 98
risk minimization 113
Rivera, Oswaldo 24, 48
Russia 61-62

Sauer, Karl 28, 124
sectoral fallow 108, 122 *see also aynokas*
Smith, Clifford 16, 18, 124
social elements 64, 99
social integument 98
soil 16, 18-19, 36, 73
  air content 19, 94
  drainage 1, 2, 55, 58
  enhancement 93, 94
  environment 91
  erosion 67, 89, 107
  fertility 46, 58, 62, 69, 78, 89, 93-94, 101, 105, 111, 116, 122
  gully 107
  humidity 69
  impoverishment 89, 92
  management 77, 91, 93, 96, 100, 112, 117
  profile 22, 27, 57
  salinity 23, 78, 17
  structure 19
  temperature 19
  type 55, 73, 7, 107, 111
  water content 57, 76, 107
state 8
  agencies *see also* FONCODES; PIWA;
  centralised 25-26, 37, 67, 82, 105
  collapse 30, 32, 121

despotic 16, 24, 29
organization 5, 14, 16, 35
suka kollus 1, 49, 53, 55, 86, 120, 123
sukallas see tara suka
suni 46, 64-65
alta 94
Superior Military College *see* University
sustainability 2, 40, 62-63, 65, 89
cultural 63-64
economic 63
sustainable development 1, 2, 40, 63, 121
sustainable land use 89, 118

tara suka 85, 108
temperature range 20, 21 *see also* climate; microclimate;
terraces 29, 33-34, 62, 106-107
abandonment of 3, 106, 109-110, 117
surface 107
walls 107
three field system 1, 97 *see also* *aynokas*
Tiwanaku 4-15, 17, 19, 24-25, 28-35, 40, 47-48, 50, 53, 83-87, 99, 109, 119, 121-122
valley 26, 48
Totora 28, 114

Uhle, Max 7, 15, 124
Umayo, Lake 63, 81
University
Agricultural College, Tiwanaku 84
Catholic 49, 50
National Agrarian 63, 95
of Chicago 47, 83
of St. Andrew (UMSA) 49
of Wisconsin 35
Superior Military College (EMI) 49
Universidad Nacional del Altiplano (UNA) 41-42, 75
Uru 9, 28, 31

Ventura Lope, Vicente 75-78, 124
vertical integration *see* integration

waru waru 1, 28, 45-46, 64, 73, 76-78, 80-81, 96-97, 119-121, 123
water harvesting 108
wetland 48, 78, 85, 87, 90, 105
Wilajawira project 48, 83
Wiñaymarka 48-49
Wittfogel, Karl 26
women 65, 72-73, 77, 79 *see also* Mothers Clubs

Yungas 110